The DISConnection:
How to Interface Computers and Video

by Gerald A. Souter

Knowledge Industry Publications, Inc.
White Plains, NY

Video Bookshelf

The DISConnection:
How to Interface Computers and Video

Library of Congress Cataloging-in-Publication Data

Souter, Gerry.
    The disconnection.

    (The Video bookshelf)
    Bibliography: p.
    Includes index.
    1. Interactive video.   2. Optical disks.   I. Title.
II. Series.
TK6643.S67   1987   621.388   87-17235
ISBN 0-86729-218-0
ISBN 0-86729-219-9 (pbk.)

Printed in the United States of America

10   9   8   7   6   5   4   3   2   1

"For my partner, Janet."

Contents

List of Figures

ACKNOWLEDGMENTS

To the following—and many more—I owe a debt of gratitude for their time, advice and encouragement:

Lynn Yeazel; Les Goldstein, Bob Sandidge, New Orient Media; Carl Renalls, 3M Corp.; Nebraska Videodisc Design and Production Group; Mitsubishi Electronics; Bruce Gjertsen, Micro Ideas; Tim Beekman, Burke Technologies; Panasonic Industrial Co.; IBM Corp.; Apple Computer Co.; Digital Techniques, Inc.; RCA Photo Library; Nicholas V. Iuppa; NCR Corp.; Smithsonian Institution; Robert Lightfoot III.

Foreword

The DISConnection was born of curiosity and necessity. By trade, I am a video producer, writer and director. My interest in interactive video was ignited by an assignment to incorporate videodiscs into the design of a "video wall" production— a production in which 18 interlocked screens were controlled by microcomputer. At that time, I was not entirely familiar with videodiscs. My first call for help was to Carl Renalls of 3M Corp. Some time before, I had seen him demonstrate how to control the selection of images with a keypad and a Pioneer disc player. The quality of the picture was superb and the ability to hop around the disc seemed magical to an old film and tape man. I was hooked. I wanted to find out more about this dynamic new medium.

I discovered that learning more is not that simple. While I found a world of people creating training programs, point-of-sale kiosks, archival databases and even entertainment, research material was scattered. I began sifting through the very few books on videodiscs and found that most covered only a single aspect of the subject: program design, directories of available discs or technical works that were beyond my immediate needs. I wanted to see the overall picture. Where did discs come from? How did the technology develop? What was the computer connection and what could I accomplish with interactive video?

This book is the result of those questions. Once the project was launched, manufacturers helped me fill my studio with computers, interfacing systems, computer software, videodiscs and a stack of manuals 3-feet high.

Before starting out, I had to define my subject because interactivity with images covers a broad spectrum of possibilities. For instance, there are interactive slide programs offered by a few production houses in which an audience can actually be polled at its seats via keypads and a menu of choices on the screen. By selecting from the menu, the audience determines what direction the program will take. When the majority decision is tabulated, the slide projectors soldier their trays around to the proper starting point and away you go. This march of the trays takes time and, while the effect is novel, the result is limited by the patience of the audience and the ability of the program to keep SOMETHING on the screen while the process takes place. Ho hum.

With the definition narrowed to MOVING images, film is eliminated by the mechanics of the film projector. This leaves us with video and the choice between linear access to information and random access to information.

Linear access involves searching for the desired information from the beginning to the end of the storage medium until the priceless kernel is located. Linear access generally pertains to magnetic tape stored on reels or cassettes. To reach "D," you must roll through "A," "B" and "C." Mainframe computers can handle linear access quite well because of their high speed, but the process is very tedious when using microcomputers or video recorders. Many of the early interactive video programs were written to work with videotape and some are still on the market. Where speed of access is not important, tape systems are viable.

The beauty of true interactive video, however, lies in its speed of response time which can only be achieved with random access—going straight to "D" by bypassing "A," "B" and "C"—coupled with videodisc technology. For this reason, I will cover only random access interactive video.

After the mechanical parameters were established, I came to understand the term "interactive videodisc system." Interactivity occurs when the videodisc, its player and a microcomputer are brought together in such a way that the viewer can ask questions, be offered choices and, in general, manipulate both text and video images. The process allows the user to determine the sequence of information through response to the program material.

I must supplement any definition of interactivity with an important reminder: Interactivity is highly subjective, there are many levels, both from the user point of view—the software—and the set of specifics imposed by the hardware. We will examine this software/hardware synergy in some depth since both elements must be in harmony for the end product to be successful.

Since my background spans both computers and video, a teaspoon of both is included along with a rather complete glossary of jargon covering both technologies. If you are fluent in either of these disciplines, you can skip to what you need. However, even readers fluent in one or the other may want to look at the glossary because—as I am discovering—fluency in either computerspeak or videospeak does not totally qualify one to solo first time out when cobbling together an interactive system. The task often seems comparable to translating the *Rubaiyat of Omar Khayyam* into Esperanto.

Finally, we will examine the elements needed to create an interactive videodisc program—a task many find intimidating. Some may be daunted by the huge sums of money involved. Producing an interactive program is not in the same league as hacking together a home movie, nor is it akin to producing an industrial videotape, despite the considerable outlay of money required for such a task.

The process of creating an interactive videodisc program is complex and requires a multitude of skills. If the task began and ended with the creation of a motivating, entertaining program on tape and the "mastering" of that program on a videodisc, I would be at home and this book would be considerably shorter. This, however, is not the case.

The first stage is the planning. What is the objective of your program? Who is your audience and what responses are you expecting to elicit? What questions will you ask? What choices will you offer? How will the user interact with your program? Will this journey through the computer/video labyrinth be worth the trip? After all, workbooks, computer-aided instruction (CAI) and point-of-sale visual aids have been around for a long time.

After committing to interactivity, computer programming is the next step. Authoring languages are available, with varying degrees of user-friendliness. One must thoroughly understand the relationship between the computer and the video-disc player. The author must interweave technologies, mutable words and immutable functions. Microprocessors have to speak to microprocessors and the result must be effective human communication.

Then comes production. Camera crews, taping equipment, logistics and editing are all tied to the requirements of videodisc mastering; these elements move on a schedule as a train moves through the night. The casual experimentation that is part of the creative process and the improvisation that is the cache of the director must be structured around a plan. The plan dominates the interactive objective. A scene shot once must often be shot again and again with different outcomes to match different user responses. The structure—to be effective—must have all its elements strictly counted.

The parts are then fused together. Images locked in frames have definite identities in an interactive program. Each frame has a number or an address– a strictly accountable reason for being. Through the unique conversation between computer and videodisc player, every image becomes part of the interactive experience.

When the computer control program is written and the video images are assembled, the result is combed clean of errors and bugs and transferred to the videodisc. The transfer process itself is a wonder of science and ingenuity.

If, by this time, you are not completely intimidated, then you can join with the brave souls who venture forth, manual clutched in sweaty hands, to carefully press buttons, attach cords and read deep into the night. Believe me, while the effort builds character, the payoff is worth all the trouble.

Remember, this technology is still in its infancy. It is a communications medium that is evolving as these words are written. This book examines specific systems and authoring languages as examples of what are available at the time of writing. Use these examples as springboards for your own investigations into the technologies current at the time of reading. The excitement of interactive video is watching its evolution and, once armed with some knowledge, selecting what fits your needs from that evolution.

We will begin with a brief immersion in history so that we understand a bit about the ground that has been trod before us. Because our subject is the combination of many technologies, an historical overview necessarily shows how developments in several fields have combined to give us interactive video.

To watch the technolgies of computers, television, videotape and videodiscs advance individually and yet join together is an interesting exercise. Take a few minutes to observe some of the genius behind our subject—the DISConnection of words and images.

1 The Evolution of Interactive Videodiscs

In this mixed bag of history, ranging from the first recorded video images and the ability to store those images to the roundabout saga of the computer, the twists and turns are considerable. We will start with the earliest attempt to record video images.

THE FIRST VIDEODISCS AND VIDEO

While it is difficult to pinpoint who really created the first video because of so many independent efforts overlapping one another, a 24-year-old German fellow seems to have inherited the mantle. His name was Paul Nipkow and his medium (providentially for our story) was the spinning disc.

The year was 1884 and Nipkow's disc utilized a mechanical scanning process. The disc contained thousands of minute holes and was placed in front of an illuminated object. The disc was spun at a fixed rate of revolutions. The holes allowed the illumination reflected off the object to pass through to a photoelectric cell as bits of light and dark. These bits were converted into electric impulses by the photo cell. Another disc, spinning at the same number of revolutions, converted the impulses back into bits of light and dark illumination on a viewing screen. (See Figure 1.1.) However crude this early effort was, the Nipkow disc endured well into the 1920s as the standard for research. Only a couple of elements were missing—the ability to transmit the electric impulses over a distance and the ability to store the images. Guglielmo Marconi would have to contribute his wireless radio and Lee Deforest, the "Audion" signal amplifier tube. (See Figures 1.2 and 1.3.)

The early 20th Century saw many advances in what a 1907 issue of *Scientific American* magazine called "Television." Two inventors, Vladimir Zworkin and Philo T. Farnsworth, eventually broke away from the mechanical scanner by developing electron vacuum tubes that "scanned" an object and reproduced the image on a photosensitive surface. The results of the scanning process could then be instantly transmitted. Because of this development, television as we know it began to evolve (from 1927 to 1930).

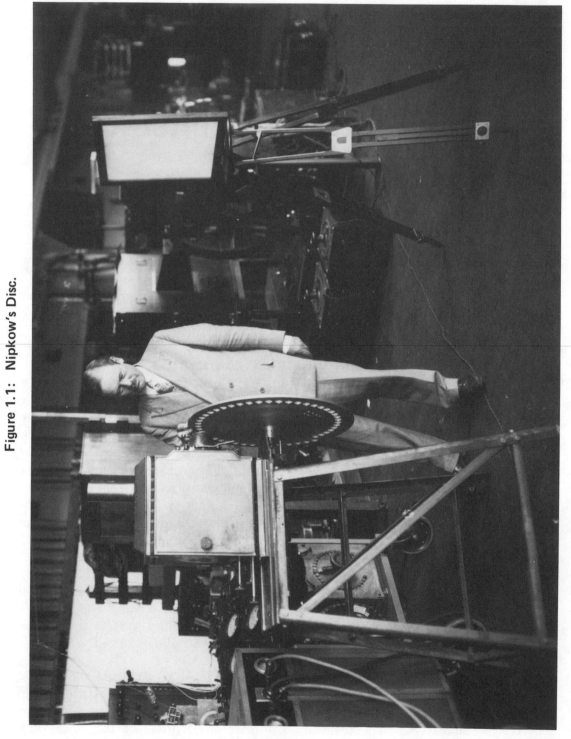

Figure 1.1: Nipkow's Disc.

Reprinted with permission of the Smithsonian Institution, Washington, DC. Photo No. 85-12139.

Figure 1.2:   A demonstration of Nipkow Disc transmission.

Reprinted with permission of the Smithsonian Institution, Washington, DC. Photo No. 85-12138.

Figure 1.3: Diagram showing how the Nipkow system worked.

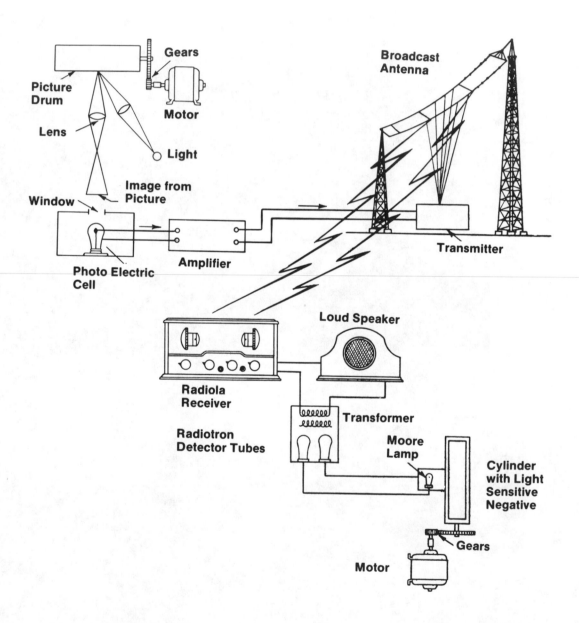

Reprinted with permission of the Smithsonian Institution, Washington, DC. Photo No. 85-12137.

The Theory of Television

Movies are made up of single pictures that are filmed at a fixed rate of 24 frames per second. The images are then projected at the same rate to achieve a "moving" picture. An illusion of actual motion is created because the human eye relies on "persistence of vision." In other words, we fill in the empty spaces mentally to create a continuous or whole image.

Television also makes use of persistence of vision. We "see" a whole picture, but what the eye is actually seeing is an organization of light, dark and colored dots.

The video or TV camera sees an illuminated object in the form of light dots, not unlike Nipkow's disc. These dots are scanned by an electron gun and projected onto a photosensitive surface. Particles on the surface convert the light into electrical impulses when scanned.

These impulses are then transmitted to the receiver and scanned onto the cathode ray tube (CRT) in your TV set, or carried by wire to your monitor in a closed-circuit configuration. An electron gun in the CRT, in a fashion similar to the camera's gun, re-scans the dots on the inside of the picture tube.

In film, each frame is a unique photograph. In video, each frame is made up of two video fields. The scanning beam moves down the inside of the tube, laying down lines of dots, 262 lines to the screen. Each line represents one-half of a line of image. The trip takes 1/60th of a second. Once the beam reaches the bottom of the screen, it returns to the top (causing the black bar you see when your picture "rolls over") This return is called the "vertical interval."

Back at the top, the beam repeats its scan, this time laying down another 262 half-lines in another 1/60th of a second. The first screen scan lays down "even" lines and the second, "odd" lines. To create a complete frame of video, these two scans are interlaced to form one 525-line picture every 1/30th of a second (1/60th + 1/60th = 1/30th of a second for both scans). This gives us 30 frames of video per second for our "persistence of vision" to unite into the illusion of moving images.

THE FIRST VIDEODISC STORAGE SYSTEM

In the 1930s, the lines of resolution were closer to 60 than 525. TV had a long way to go. By this time, however, amateur television enthusiasts had a jump on everyone else, at least in the British Isles. The Nipkow disc led to a disc of another kind—the first videodisc storage system.

The Baird Televisor

James Logie Baird started experimenting with mechanical image scanners in 1923 and, by 1926, he was ready for a public demonstration of his Baird Televisor (see Figure 1.4). The videodisc was a sideshow to the main event, but it outlasted its humble objective.

Baird used a Nipkow disc scanner at the transmitter to scan the subject. His Televisor employed a similar disc and a neon bulb that was set up on the receiving

Figure 1.4: Hand-built prototype Baird Televisor.

Photo courtesy of the Independent Broadcasting Authority, London.

end. In 1926, the BBC arranged a series of half-hour experimental transmissions with the Baird Television Co. A viewer had to have the Televisor set up and running prior to the broadcast so the Nipkow disc would be spinning at the same speed as the scanning disc at the transmitter. To ensure that the Televisor was properly set up so the viewer wouldn't miss the program, Baird came up with a system of inscribing video signals on a wax "Phonovision" record.

This was possible because the resolution of those early transmissions was only 30 lines per screen, or 12.5 video frames per second.

The early images were pretty crude and except for these experiments, the video-disc of 1926 was doomed to the back shelf in the closet. Its like was not to materialize again until the early 1960s.

EARLY COMPUTERS

Where were computers at this time? About the time Herr Nipkow was drilling holes in his disc, computers were called Arithometers or adding machines. In 1888 a young man named Burroughs took out patent #388,116 for a Calculating Machine. By 1889, he had built 50 machines. By 1907, 50,000 had moved out of his factory.

In 1924, the punch card using a rectangular hole was the product of the Electric Tabulating Co. headed by Herman Hollerith. He and his engineers also incorporated telephone technology into their punchcard machine systems. Telephone switchboardlike plug panels were used to transmit information from any column in a punch card to any register in a bank of sorting and integrating machines. This plugboard system is still in use today.

Another significant event occurred in 1924. A gentleman named Thomas J. Watson posted a sign on the walls of his factory, the Computing-Tabulating-Recording Co., that said, simply, "THINK." He also changed the name of his successful operation to International Business Machines Corp. IBM took its first steps toward an uncertain future.

Those early machines were still mechanical. It took a World War and the technology that followed to bring computers into the electronic age.

The Technology of World War II

The atmosphere of the World War II years gave impetus to important developments in early computer technology; some of the most memorable are a German code device and its British decoding counterpart, the first programmable digital computer and vacuum tubes.

During the 1930s, the Germans settled on what they felt was a totally secure encoding/decoding device—"Enigma." The machine was used for all radio transmissions of secret information just prior to and during World War II. It used a combination of multiple rotors, each engraved with letters of the alphabet. As the rotors rotated, combinations of letters were produced that could be decoded into the original words. Key code setups were changed weekly, or even daily. As long as the same setup code was used by both the transmitting machine and the receiving machine, messages were transmitted.

Polish mathematicians managed to build some replicas of the machine and smuggled them into England. Using the replicas as a foundation, a group of British mathematicians built a decoding device that was capable of taking encoded messages intercepted by British Intelligence and running a set of possible permutations until the message was decoded. The device used mechanical means: belts, pulleys, wheels and numerous sets of Enigma-style rotors. The British called the machine the "Bronze Goddess" and its information was disseminated to Allied armies throughout the war under the cover name "Ultra."

While this mechanical computer was listening in on the Nazis, Thomas J. Watson of IBM was busy looking for new ways to expand his calculating machine company. To further that goal, three engineers—Clair D. Lake, Benjamin Durfee and Francis E. Hamilton—who had been assigned to produce an Automatic Sequence Controlled Calculator came up with the Mark 1. (See Figure 1.5.)

This device used mechanical relays but was the first programmable, digital computer. It weighed 5 tons and used punch card stock, paper tape and manually controlled switches. The Mark 1 was delivered to Harvard in 1944 and was set to computing ballistic tables for the U.S. Navy. Although it was obsolete from the time it was switched on—because of the invention of the vacuum tube—the relay-clattering monster worked until it was scrapped in 1959.

Figure 1.5:   The IBM Mark 1 Automatic Sequence Controlled Calculator.

Photo courtesy IBM Archives.

Vacuum tubes were first used effectively during the war in the form of an anti-aircraft gun—the M-9 Gun Director. In this device, vacuum tubes changed electric voltages to mechanical signals that controlled the servo motors which in turn moved the gun.

Post-War Developments

ENIAC

The late 1940s saw a flurry of activity in the computer field. While the Mark 1 was in use at Harvard, J. Presper Ekhart and John W. Mauchly made use of the vacuum tube's switching speed over the mechanical relay to produce ENIAC.

This Electronic Numerical Integrator And Computer was unveiled in 1946 and filled an entire room. It could perform 300 multiplications per second compared to the one multiplication-per-second rate of mechanical computers. These computations were made at the speed of 100,000 pulses per second, which was a phenomenal rate at the time. Today's home computer, by comparison, operates at a rate of some 4,000,000 pulses per second.

ENIAC weighed 30 tons and employed 18,000 vacuum tubes plugged into 40 modules that plugged into each other. Unfortunately, all the wires and cables connecting its electronic viscera had to be rewired each time a new program was required. Another problem was heat caused by ENIAC's vacuum tubes, coils, resistors and assorted electronics. As with its predecessors, ENIAC was assigned to ballistic tables but also found useful application in the fields of meteorology, nuclear energy, cosmic ray studies and aerodynamics. It operated continuously until 1950 and now resides, a museum relic, in the Moore School in Philadelphia.

An offshoot of this growth of computer technology was the programmer—the person who told these early computers what to do, replugged those wires and turned all those hundreds of switches. This was before the days of computer languages, so the complex job was left in the hands of the mathematicians.

Computer development following ENIAC is marked by a bewildering barrage of alphabet soup: EDVAC, EDSAC, SEAC, SWAC, MADM and even MANIAC. Three computers—SSEC, BINAC and UNIVAC—were especially significant during this time of rapid growth.

SSEC

In 1948, IBM unveiled the Selective Sequence Electronic Calculator (SSEC). This was the first computer able to store its own operating instructions as though they were data so they would not have to be laboriously reloaded every time the machine was fired up. It made extensive use of vacuum tubes and was considered such a success by IBM that it was displayed behind glass walls in its New York City World Headquarters where passersby on Madison Avenue could admire the technology. (See Figure 1.6.)

Actually, the SSEC was a mixed bag. It used 12,500 vacuum tubes combined with 21,400 electromechanical relays and stored data on paper tape. Some criticized its usefulness because of the number of vacuum tubes it used. Vacuum tubes of the period had a working life of about 3000 hours. Allowing for the 12,500 tubes in the SSEC, the laws of probability dictated that one tube would fail about every 15 minutes. Since it took about 15 minutes to find a blown tube, some felt the SSEC was nothing more than a huge, useless, vacuum-tube recycling machine.

The SSEC did not have only critics, however. An exacting market study carried out in 1949 showed that there was no future in the computer market because the nine SSECs then operating could do all the work.

BINAC

Despite this gloomy prediction, the Messrs. Eckert and Mauchly of ENIAC fame started up the Eckert and Mauchly Computer Co. and, in 1949, produced the BINAC. Amidst the flurry of computers appearing on the market, BINAC was unique in one major aspect—it used the binary number system which eventually became the industry standard.

The binary system requires only two numbers—a one and a zero. According to binary, something is either "on" (1) or "off" (zero); "open" (zero) or "closed" (1). As with the punch cards of the time, there is either a hole in the card or there is no

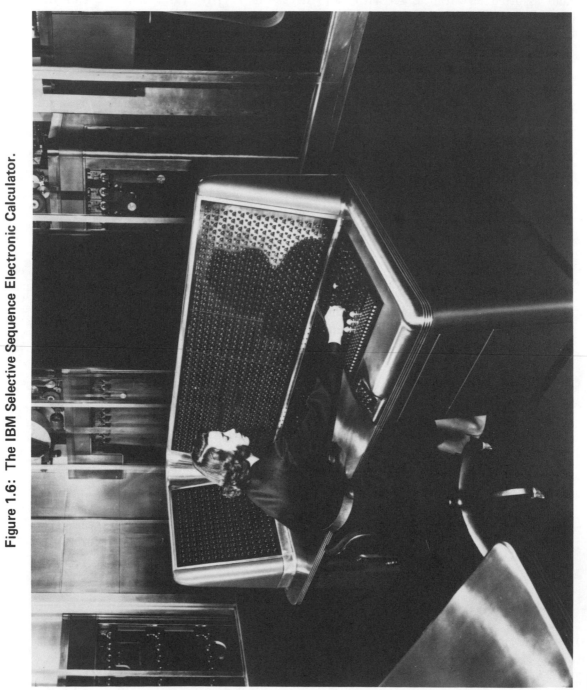

Figure 1.6: The IBM Selective Sequence Electronic Calculator.

Photo courtesy IBM Archives.

hole in the card. It is a Base-2 system as compared to our decimal system which is Base-10.

In binary code, each number is called a "bit" and, for example, 8 bits equal one binary "word," or a "byte." This code is used to interpret letters of the alphabet or numbers into sequences of 1s and 0s that can be read by the computer's operating system. This code is now recognized as an industry standard (and will be dealt with later).

The significance of the BINAC was that it led Eckert and Mauchly to the next great step in the evolution of computers—the Universal Automatic Computer, or UNIVAC.

UNIVAC

This 16,000-pound machine had only about 5000 vacuum tubes and *no* mechanical relays and, therefore, was the first truly *electronic* computer. It was put to work right away by the Remington Rand Corp., which acquired the Eckert and Mauchly Computer Co. UNIVAC went on to compute the 1950 U.S. Census and achieved media coverage when it helped CBS News predict the outcome of the 1952 Presidential election. It suggested that Dwight Eisenhower would defeat Adlai Stevenson by 438 Electoral College votes to 93. When the dust settled, Eisenhower was President by a margin of 442 votes to Stevenson's 89.

After that election, public awareness of computer power was established. From that point on, the machines became smaller and more powerful, but the future of computer science was not imagined.

UNIVAC also introduced an innovation that allows a rather nice transition into our parallel history of video. The computer was first to use magnetic tape for storing data. It differed from the tape developed by Telefunken some years before (discussed in the next section) because it was iron-plated metal (the forerunner of today's metal tape).

AUDIOTAPE AND VIDEOTAPE

In addition to the coding device "Enigma," the Germans made another significant technological contribution during World War II—the creation of an electromagnetic means of storing audio information in the form of iron oxide particles coated on a plastic tape. This pioneering effort of the German firm, Telefunken, fell into Allied hands after the war through the efforts of John Mullin, who later became a consultant for the motor manufacturer, Ampex.

Shortly after the war, the first American tape recorder was introduced. The stage was set for following the storage of taped sound with the storage of taped images.

To achieve that end, John Mullin was hired by Bing Crosby Production's Electronic Division to develop image storage on tape, using audiotape principles. Early experiments were not a great success.

They tried passing the tape over a record head at 120 inches per second and the results were ghastly. The major obstacle was the amount of information required to store video as opposed to audio. It was like trying to stuff 50 pounds of potatoes into a 10-pound bag. In electronic terms, this disparity is interpreted in terms of bandwidth. You need more bandwidth for video storage than for audio storage.

The Crosby-Mullin team attempted to run the tape at 120 inches per second and then split the signal onto 10 parallel tracks. This splitting altered the effective speed to 1000 inches per second but the pictures were still terrible. An example of the standard they were seeking would be the current U.S. broadcast standard that requires the smallest element broadcast be 1/336th of the height of the picture. The best the Crosby-Mullin team could manage was 1/35th the height of the picture. It is easy to understand why the early 1950s became the golden age of live television: there wasn't any alternative.

While Mullin and Crosby were scratching their heads, engineers at Ampex were trying to use FM (frequency modulation) signal encoding techniques to reduce the amount of information required for video signal storage.

As it turned out, the best way to whittle down this information was to reduce the number of octaves required. An octave is the frequency span from a single frequency to twice that frequency. Video uses frequencies from 30 Hz (thousands of cycles) to 4.2 MHz (millions of cycles) sprawled across 18 octaves. FM radio, on the other hand, requires only eight octaves. By transposing the video signal upward to cover the frequency range between 4 MHz and 8 MHz, it fits into fewer octaves and makes it easier to control frequency response over the 18-octave range. The process was called "equalization" and led to *the* major step in video image storage: Quadruplex recording.

The solution, developed by Ampex engineer Charles Ginsberg, was to use four recording heads spinning at right angles to the tape. Though the tape moves at a rate of only 15 inches per second, the spinning record/playback heads create an effective writing/playback speed of 1560 inches per second spanning four octaves. Not long after this discovery, Ampex videotape recorders, using tape 2 inches wide, became the industry broadcast standard. The year was 1956.

By 1960, helical tape recorders had evolved. These machines wrapped the tape diagonally around a drum spinning in the opposite direction. The helix formed by the tape path allowed video information to be written in diagonal lines and the tape speed was slowed down to 9 1/2 inches per second. Tape width was reduced from 2 inches to 1 inch and the steady march toward the modern home cassette recorder had begun.

EARLY VIDEODISC RESEARCH

As an aside to this heady advance of video technology, videodiscs popped up once again in 1965. An outfit called Magnetic Video Recording made sports fans delirious with joy and helped coin the phrase "instant replay." The device was the HS-100, a disc coated with magnetic media that spun at 1800 RPM and could record an event on tracks much like it could be recorded on a music record. The exciting part was that the event could be recreated in slow motion by holding the playback head over a single track and freezing, or stop-framing, the event. Football fans were ecstatic, referees were terrified and the TV sports industry just smiled and counted the money.

The huge strides in videotape technology were still hindered by one major difficulty—duplication of tapes. It still took an hour for a duplicating tape machine to duplicate an hour-long tape produced on the master machine. Researchers looked at the way music LPs were stamped from a master disc created from the original audiotape.

One of the first attempts to try a similar process with video came from a merger that would have been unlikely 20 years earlier. Germany's Telefunken (of audiotape fame) joined with Britain's Decca to create TelDec. The idea was to reproduce videodiscs using the same track and stylus scheme established by the music industry.

TelDec faced the same problems as its fellow videotape developers. Video images require about 210 times more bandwidth than audio recordings. A 20-minute music record loaded with video information would play for only five seconds. Like the Crosby-Mullin team before it, TelDec tried to speed up the process. It spun the record at 1800 RPM with 280 grooves per millimeter. The needle raced along over each groove, chattering across a series of bumps, and since each bump caused a voltage fluctuation TelDec was using the same principle as FM encoding.

FM Encoding

FM carries no information other than highs or lows, which means it can be reduced to the level of Morse Code—a series of dots and dashes. This level allows four variables: a dot might indicate a signal above a certain amplitude while a dash might mean a signal below a certain amplitude, and the absence of either a dot or a dash can also be read as a variation in the signal. Dots and dashes can be converted into bumps (as with TelDec) or pits (our current videodisc technology) and the frequency modulation can be transmitted and converted into sound and/or pictures. With TelDec, each 8-inch disc lasted only 10 minutes. The partnership eventually folded but its contribution of FM encoding techniques for videodiscs was lasting.

The CED Disc

Systems requiring a stylus to actually touch the surface of the spinning disc reached their commercial nadir in the form of RCA's capacitive electronic disc (CED) product, which achieved wide acceptance before being withdrawn in 1984. This system made use of a characteristic of electricity called "capacity."

The ability of a device to store a charge of electricity is called its capacity. A capacitor is used in radios to select the frequency desired. It is sensitive to frequencies, allowing higher frequencies to pass through more easily than lower ones. The CED disc made use of this capacity theory by making the record and the stylus form a capacitor. The record was etched with pits much like the bumps on the TelDec discs. As the stylus rode over the pits, the capacity of the combination stylus-disc capacitor varied, tuning an electronic circuit that permitted reconstruction of the recorded signal.

The RCA disc rotated at 450 RPM and was mastered much like audiodiscs. The grooves were extremely narrow, permitting about 38 to fit in a single music LP groove. The pits were roughly .0002-inch in size, or half a micron.

RCA figured that the use of off-the-shelf technology would allow it to market its players and discs at a price well below the optical disc systems of the period. But it had not reckoned with the explosion in videotape recorder and cassette sales. RCA had pushed stylus discs about as far as possible, but the market was not solid enough to support two disc systems, especially since disc players could not record material off the air or from cameras as videocassette recorder systems could.

The other problem was information access. A CED disc was a linear access device, especially in the RCA format where the user could not get at the stylus to reposition it. Therefore, almost all the titles for CEDs were movies or educational programs that could only be started and stopped. In 1984, RCA dropped the entire line as a dead-end venture.

INTEGRATED CIRCUIT RESEARCH

An event occurred on December 23, 1947 that figures prominently in our brief look at the history of both videodisc and computer hardware: William Shockley, John Bardeen and Walter Brattain invented the transistor. Until that invention, computers were held captive by the inefficiencies of the vacuum tube.

A transistor is a part of a circuit that can turn the flow of current on or off in the same fashion as the tube. However, there is no vacuum, no heat and nothing to burn out. The transistor performs its functions enclosed in a block of semiconductor material such as silicon. This encapsulation has another big advantage—speed. At the time of its invention, even the most primitive model could perform the switching function 20 times faster than a vacuum tube.

Until the transistor, circuit designers were thinking in terms of room-size devices for computing. After the transistor, designers imagined new devices that could utilize circuits composed of from 50 to hundreds of thousands of these tiny switches. This is where the major snag arose. Sure, you could cram thousands of these mini-components into a design that would outperform anything ever dreamed of before. But who would wire all that mess together? A circuit containing as many as 100,000 components would require a million or so *hand-wired* connections.

The first completely transistorized computer, the Control Data 1604, was designed to contain 25,000 transistors, 100,000 diodes and baskets full of resistors and capacitors. It languished as this "tyranny of numbers" was reckoned with, because one bad connection in the constant flow of data signals spelled failure for the system.

Another problem presented to designers was the fact that all the wire needed to create these huge circuits undermined the chief advantage of the transistor—its switching speed. A device that could switch at the rate of one time a second was useless for computers. Raising the count to 1 million times makes the computer a useful tool, but at a rate of 1 billion switches a second, or one step of the program being completed every nanosecond, the machine creates what we have today—a revolution.

This switching speed is necessary because of the virtual stupidity of the computer. Since it reduces all tasks to the binary or Base-2 numbers concept—one or zero, on or off—it takes a dozen computations just to alert the machine that someone is hitting its keys.

In 1958, an engineer working for a small firm, Texas Instruments, had an idea that has become known as the Monolithic Idea. Jack Kilby suggested that all the components—transistors, resistors, diodes—that make up a circuit could be incorporated on the same chunk of silicon. He sketched out his idea, then persuaded his bosses to let him try and build the thing. On September 12, 1958 he recorded in his log that this first, crude, circuit-on-a-chip worked the first time power was applied.

Meanwhile, in another part of the country, two engineers named Bob Noyce and Jean Hoerni had the same idea—dubbed the "Planar Process" by their company, Fairchild Electronics—and they took the problem one step further. Jean Hoerni's idea was to protect the delicate transistor by placing a layer of silicon oxide on top of the transistor chip. This protection would keep the transistor free of contaminants and seal the connections.

Noyce examined Hoerni's idea a bit more closely and stumbled on the idea that could solve a major problem. Wiring transistors together was a great pain, but with this coating of silicon oxide on top of the layered silicon transistor, a wire could be poked down through the oxide to an exact connection on the chip. As the idea developed, an even bolder thought formed: Why use wires at all? Why not photoprint the connecting lines on the flat surface? Why not try to connect more than one transistor on a single chip? Why not connect a number of components, using the printed lines, and form an *integrated* circuit on a single chip of silicon?

Just six months after Jack Kilby developed his idea for this integrated circuit, Bob Noyce also produced a solution for the Monolithic Idea.

On March 24, 1959, the integrated circuit appeared for the first time at the annual meeting of the Institute of Radio Engineers. This invention that would be set alongside the electric light, the telephone and the internal combustion engine as a major turning point in our civilization was almost completely ignored by the experts. Texas Instruments and Fairchild became locked in a legal brouhaha over who was entitled to the patent on the device. In the end, the two companies struck a joint licensing deal. Jack Kilby was credited with the first discovery of the circuit while Bob Noyce became the father of the printed integrated circuit that eventually evolved. Everyone made a ton of money and today, the industry is worth billions each year.

The 1960s and early 1970s saw the shaping of the computer into the ubiquitous machine we know today. The huge mainframe computers of the 1940s and 1950s shrank in size and doubled in memory. The "minicomputer" evolved and proliferated on college campuses and in small businesses where the mainframe was beyond the capacity of the pocketbooks.

DEVELOPMENT OF THE MICROCOMPUTER

This proliferation afforded computer access to a wider sector of users. These neophytes quickly developed a taste for this new computing power and waited for the chance to have machines of their own. Those who cobbled together their own crude devices shared their plans with other electronics-oriented folk. A computer underground developed, challenging the annointed data processing experts who had control of the mainframes and minis within the college and corporate hierarchies.

These engineers, students, hobbyists and basement inventors simply wanted something a bit more personal.

Personal Computers

Intel's 4004

In 1971, an engineer named Marcian E. "Ted" Hoff worked for a small company named Intel (Integrated Electronics Corp.), which was co-founded by Bob Noyce in 1968. Hoff managed to cram all of the circuitry needed to operate a calculator onto a chip of silicon 1/6-inch long and 1/8-inch wide. It was called the 4004 and was the first microcomputer.

Because no one in mainframe data processing knew quite what to do with these radical chips, Intel was stuck with many unsold 4004 microprocessors. In order to unload the inventory, it put the orphan in its catalog at a rock-bottom price. Hobbyists and electronics buffs leaped at the opportunity and bought up the chips. The 4004 became a huge success and all across the country, basement computers were being cobbled together.

The Mark-8

True computing was still in the hands of the data processing experts, and at about the same time that the 4004 became successful IBM announced the System/32. This was a computer the size of an average desk that contained all the necessary computer hardware.

Inspired by the success of the 4004, however, Intel expanded its limited 4-bit operating system to a faster 8-bit system. In 1974, the magazine *Radio Electronics* published a set of plans to build the Mark-8 Personal MiniComputer using the Intel 8008 microprocessor.

The Altair 8080

The momentum had taken hold and that same year two new chips arrived: Intel's 8080 and a competitor designed by two former Intel engineers, the Z-80 by Zilog. The 8080 microprocessor showed up in the pages of the January 1975 issue of *Popular Electronics* magazine as the heart of the first personal computer that became widely available. The Altair was built by MITS (Micro Instrumentation and Telemetry Systems).

The Altair bore no resemblance to current microcomputers. There was, for instance, no keyboard—raw, binary code could be loaded into the machine by flipping switches. A simple instruction such as STOP took eight switch flips. Toggling in a program that was several hundred bytes long meant more than a thousand switch flips and, if only one binary digit (a "1", or "0") was wrong, then the machine just sat there—or worse, went quite mad, messing up the bits that had been entered correctly. Another grim fact was that the information entered was all in the Altair's RAM (Random Access Memory). If you switched off the power, all that switch flipped information evaporated and the program would have to be re-entered—a process that took hours.

Eventually, devices such as Cromemco's Byte Saver introduced the EPROM to computer enthusiasts. This device allowed a ROM (Read Only Memory) chip to be imprinted or programmed with operating instructions that would not be erased when the machine was shut down.

If these limitations were not enough to make a person weep, there was also no way to store programs except by the age-old method employed by early computer pioneers—paper tape. To save some people's sanity, a high-level language had been developed called BASIC (Beginner's All-purpose Symbolic Instruction Code). This language allowed instructions to be typed in English symbols and words. These symbols were then interpreted by software into binary for the computer to execute. In 1975, a small company named Microsoft rewrote this BASIC language for the Altair and life for computer users became much simpler. Unfortunately for the Altair, the business acumen of its founders was not up to their ingenuity and the company went broke, but not before its build-it-yourself kits had inspired other inventors.

The Apple II

In 1976, an electronics whiz took a design for a microcomputer to his boss at Hewlett-Packard. The boss was uninterested. So Steve Wozniak and his friend Steve Jobs set up their own company, first in Wozniak's bedroom in Sunnyvale, CA, and then out in the garage. Jobs had spent a summer working in an orchard, so the name "Apple" was selected for their homespun effort. It didn't look like much, just a board and a bunch of chips and transistors that were designed to hook up to a TV set. The design was first demonstrated to some of Wozniak's fellow computer club members, then offered as a kit. Wozniak and Jobs got offers for 50 kits. When their business seemed to be growing beyond their capability to handle it, they brought in A. C. "Mike" Markkula, Jr. to help set a course for this new marketplace.

Right away, they began some market research and discovered their compilation of parts would sell much better and to a wider audience if it was self-contained and pre-assembled. Some venture capitalists had a similar idea and the Apple II was born. (See Figure 1.7.) Using an ordinary audiocassette player-recorder as a data storage system, the Apple II was a hugh success. Later, a disk operating system (DOS 3) was created by Wozniak and the "floppy disk" invented by IBM in 1971 was adopted.

Figure 1.7b: The Apple IIe microcomputer.

Photo courtesy of Apple Computer, Inc.

Figure 1.7a: The first Apple computer, designed by Steve Wozniak.

Photo courtesy of Margaret Wozniak.

To try and recount the escalation of Apple from garage to billion-dollar corporation would be both tedious and redundant in the extreme. It is only necessary to say that 1977 seemed to be the kickoff year for companies such as Apple, Commodore and Radio Shack. And while they grew and prospered, they kept looking over their shoulders at IBM, as its corporate analysts calmly surveyed the marketplace. The pioneer microcomputer makers didn't have long to wait. In 1981, the IBM PC descended from on high.

The IBM PC

It wasn't a bombshell as far as new technology was concerned. The IBM PC was a very conventional desktop computer with a keyboard, up to two disk drives, a monitor and a very ordinary—and even somewhat archaic—8088 microprocessor. The Intel microprocessor is an 8-bit device that emulates a 16-bit processor. Nothing grand, but it was a notch higher than anything else in the 8-bit world of 1981. As of this writing, those earlier chips have been replaced by a new generation of microprocessors, the Motorola 68000 series and the Intel 80286 and 80386.

People who peered inside that first IBM microcomputer case saw chips that were common to other computers. They handled 5 1/4-inch floppy disks that the other computers used. Even the operating language, MS-DOS from Microsoft, resembled CP/M, a language that had been kicking around as an industry standard for more than five years. What caused the industry to reel in its tracks and move inexorably toward this particular personal computer was the name on the package—Thomas Watson's IBM. And as we'll see in Chapter 3, this momentum also had an effect on the development of interactive disc authoring systems.

This history of computers has left some gaping holes, but it would be impossible to include everything and still have some paper left over to discuss interactive videodiscs.

OPTICAL VIDEODISCS

With our history of computers behind us, it's time to find out where videodiscs came from. Work on the theory of optical, random access videodiscs began while television was still struggling with developments such as the diffusion of compatible color images among the masses and the design of the first NBC peacock.

One research path began in frozen Minneapolis, 1961. Three engineers—Wayne Johnson, Paul Gregg and Dean DeMoss—were studying magnetic tape recording for 3M in an attempt to discover an inexpensive way to reproduce videotape recordings. Casting about for inspiration, they trod on old ground. Using an electron microscope focused on the surface of an ordinary phonograph record, they discovered the surface was smooth enough to allow the necessary high resolution required to reproduce video signal bandwidth. As researchers before them had found, the disc was the most logical solution for replicating video information (and data as well, the three noted, but their job was aimed at the home TV player), and so they pressed on.

The greatest hurdle to acceptable videodisc reproduction was to find a record and playback method that could resolve the micron-sized bits of video information jammed together on the disc surface. They tried X-rays and the stylus method that would eventually blossom, fade and rise again under the aegis of TelDec, RCA and JVC. Finally they took their problem to the Stanford Research Institute (SRI International today) for assistance. They established the following criteria:

1. The picture element size would be 1 micron.
2. Space between tracks would be 2 microns.
3. The bandwidth would be commercial broadcast standard.
4. Playing time should be 10 to 15 minutes per side.
5. Mastering would utilize electron beam machining.
6. The method of replication would be similar to that used for LP phonograph records.
7. Playback systems would be investigated. One idea was to play the record in a vacuum and use an electron beam to trace the spiral tracks. Another was to use absorption of X-rays.

The project was begun on July 1, 1961 and by August of that year the report from SRI ruled out both schemes in favor of optical readout, and record pressing was abandoned in favor of a photographic plate. The latter decision was a practical one since no record pressing facilities were available to the researchers.

They then faced the same problem that plagued Nipkow's rotating disc, but on a greater technological scale. How would light pass through a hole 1 micron in diameter at a rate of 4 million cycles per second (4 MHz) and be detected efficiently by 1961 state-of-the-art photodetectors? Using a dilute mixture of 1 micron diameter latex spheres floating in distilled water, they let the water evaporate under a microscope, leaving behind a dried honeycomb of micron-sized holes. By shining a light through the holes, they found a photodetector could indeed detect any slight off-center shift of the light source.

Next came tests to see if a stream of these micro bits could produce a useful signal at the 4-MHz rate. Using the old tried-and-true rotating disc and a 100-watt, high-pressure mercury arc lamp, they succeeded in detecting a more than adequate output signal. They were encouraged to continue their research and the last of these preliminary tests was wrapped up during February of 1962. The revolving disc, considerably modified, was able to display a 125-line video image of a black-and-white photograph. These results showed that available technology was up to the task.

It has to be understood that gas lasers were not yet available, so the researchers were limited to short-arc mercury and xenon lamps. They continued working with reproductions of high-resolution, black-and-white images and in October of 1962 succeeded in recording and playing back these images from the first optical videodisc. By June 1963, they had recorded numerous 16mm films on discs.

In all, by 1965, some 20 patents had been awarded to 3M, covering innovations in all aspects of optical disc research. Its work was halted only by the unavailability of gas lasers.

With the 1970s came the work of Philips, MCA and Thomson-CSF using the new lasers to record discs. A number of laser disc recorders were built at 3M and a durable plastic coating was devised that could be cast and cured rapidly. By 1981, 3M was recording and replicating discs in agreement with both Philips and Thomson-CSF. The optical disc had arrived.

The Marketplace

The end of the 1970s and the early 1980s saw the birth of the consumer video-disc market. One of the real problems with sorting out the who's who of this market-place was that everyone seemed to be owned by someone else.

In 1978, for example, the press was assured that videodisc players would be available in time for Christmas. The companies making the promises were Magnavox, which was actually a subsidiary of North American Philips, which was owned by N.V. Philips in the Netherlands and the Music Corporation of America (MCA), which owned Universal Pictures. Philips/Magnavox provided the hardware while MCA/Universal provided some 200 movie and other titles—at least in the catalog. The players sold out, but the discs were another matter. Converting feature films into high-quality videodiscs was a new and laborious process. The best MCA could manage was a 45-day turnaround from receipt of the program to shipping the discs.

To compound the problem, in 1979, RCA announced its SelectaVision CED disc system discussed earlier, which sold for considerably less than the optical products.

Eventually, keeping track of the players required a score card. MCA/Philips was courted by IBM and together they became DiscoVision Associates. IBM figured it could use its computer technology to create "smart" videodisc players.

In 1980, MCA/Philips/IBM added Pioneer Electronic Corp. of Japan to their camp as Pioneer announced an optical player into the U.S. market. Pioneer was already building a DiscoVision industrial player for Universal-Pioneer—the PR 7820. These players built by Magnavox (Magnavision) and Pioneer were not totally feature-compatible, so the public was faced with companies competing with themselves as well as with industry giants such as RCA which produced SelectaVision—a totally different disc format. All this time, optical videodisc movie titles were just trickling into the marketplace. IBM finally threw up its hands and went back to doing what it did best—developing that personal computer.

The videodisc marketplace during the early years from 1981 to 1983 was domi-nated by the battle for consumer dollars between the manufacturers of laser optical discs and RCA's CED disc. There was a pricing irony that became ludicrous—as the players became more sophisticated and added consumer features, they had to be sold at lower prices to keep pace with flagging over-the-counter interest. RCA and DiscoVision Associates were pumping out more and more movie and special-interest titles.

At the June 1982 Consumer Electronics Show Sony announced the first personal computer that was designed to work with video equipment, the SMC-70G.

(See Figure 1.8.) The machine also demonstrated the first mainstream use of 3 1/2-inch micro-floppy discs. This video interfacing capability would keep the Sony machine among the survivors during the following years while other less specialized computers lost out.

Another debut at the 1982 Summer Consumer Electronics Show was the Compact Disc (CD) for audio reproduction introduced by North American Philips. This laser-read product achieved almost immediate public support. Laser technology had finally found a consumer home.

By the end of 1982, Pioneer had brought out a successor to the PR 7820—the PR8210. (See Figure 1.9.) It featured entry-level interactive capabilities including a fully featured, infrared, wireless remote keypad, heavy-duty industrial-quality motors and built-in CX noise reduction. For interactive input, a special computer connection was also included. In September of 1983, Pioneer made the total commitment to interactive video by introducing the LD-V1000. This videodisc player could be controlled only by an external microprocessor, generally in the form of a micro-computer. The LD-V1000 was also suited to multi-player programs controlled by either mini or microcomputers.

While the marketplace endured its ups and downs, interactive applications were finding homes.

**Figure 1.8: The Sony SMC-70G personal computer, designed
to work with video equipment.**

Photo courtesy of Sony Corp.

Figure 1.9:   The Pioneer Laserdisc Player PR8210.

Photo courtesy Pioneer Video, Inc.

Applications

As early as November 1980, the Museum of Fine Arts in Boston pressed a video-disc created in-house that documented the museum's art collection. The disc was played on the PR7820, using only the capacity of the player's microprocessor. In its final form, the disc held 2000 photographic images of the 108,000-image storage potential.

ComCorps in Washington, DC, transferred the original slides to rolls of 35mm Ektachrome Duplicating Film and DiscoVision Associates converted the slides to tape and the tape to a mastered disc. Two short computer programs were placed on the disc to be "dumped" into the PR7820's memory. Categories and secondary categories from the art collection could be accessed by frame number and, after the last image was selected (search time 1.5 seconds), the user was returned to the secondary category menu. This disc helped validate the laser disc's capability as an archival storage medium.

Other interactive uses were being investigated as well. On March 31, 1981, a 22-month project conducted by WICAT, Inc., an interactive video developer, reported on the results obtained using interactive videodiscs in a science curriculum. The project was funded by the National Science Foundation. The report read, in part: "... the delivery of science instruction via intelligent videodisc has proven not only feasible, but superior in some situations to traditional college science instruction. ..."*

*Rod Daynes and Beverly Butler, eds. *The Videodisc Book* (New York/John Wiley and Sons, 1984).

The system used a computer (The PASCAL Microengine built by Western Digital), an optical videodisc player (DiscoVision Industrial/Educational Player) and an interface built by Texas Instruments. Instruction was given to students at Salt Lake City's Brigham Young University and Brookhaven Community College. The course material selected for the test was on DNA structure and function and the transcription and translation phases of protein synthesis. Students responded favorably to the videodisc learning experience. They particularly liked the control that allowed them to proceed at their own pace and the combination of live video and graphics images.

Another successful application reported in 1981 was the Video Patsearch designed by Pergamon Infoline. This system used a microcomputer to access the Patsearch database on the BRS (Bibliographic Retrieval Services, Inc.) network which was, in turn, accessed by users via telephone modem and personal computer. The Patsearch database contained virtually all the patents assigned since 1971, some 720,000 at the time of the 1981 report. Eight videodiscs contained all the information by patent number on 15 sides. The player chosen was the ubiquitous PR7820-2. Users subscribed to the service by the hour online and through an annual subscription cost.

According to the Canadian Patent Office where the system was being tested, this application resulted in a major breakthrough in patent searching.

CONCLUSION

By 1983, the interactive videodisc marketplace had asserted itself even though consumer videodisc sales had begun to plummet. Sales and rentals of videocassettes and recorders had eclipsed the consumer market along with cable television.

Though many pioneering companies and developers were producing interactive video programs, the TALMIS projection of 1981 held true. This computer industry research organization had determined that the corporate and institutional marketplaces were largely ignorant of interactive video and there were few established technical standards. Since the history of videodiscs was liberally punctuated with failures, dead ends and poor market performances, user apathy was understandable.

The task of moving interactive video forward would fall to the independent producers to press for new system designs, new applications and new interfacing products. The big companies would provide the basic players and raw discs, but it would be up to the entrepreneurs to cobble everything together—and that's basically what happened. Even today, the 1981 TALMIS projection is valid, but there are many more developers and choices to offset the lack of industry standards.

2 The Role of Computers

The chronology presented in Chapter 1 was intended to provide enough history so that we could understand our progenitors. You can see how the great discoveries seemed to intertwine and how we ended up with, essentially, a "top down" videodisc industry (i.e., players and discs flow from the founts of a few big companies). The computer industry—at least the microcomputer industry—has developed the opposite, or "bottom up" scheme. Anyone who can afford a computer and the time to learn a programming language can create software. The majority of software, or application programs that do the actual work, is written by independent developers.

One parallel between the computer and videodisc camps does remain, however: large companies rule. Today, Pioneer, Sony, and a smattering of others manufacture disc players. The computer industry can be described in terms of IBM, IBM compatibles, Apple and everyone else. Although computer pioneers such as Sony and most recently IBM work toward more flexible player controllers, the interactive user today has few choices among computers.

Videodisc players range from basic players with small command sets to complex machines with microprocessor controlled functions, but, essentially, they do one thing: play discs. A microcomputer can control videodisc players, balance checkbooks, graph stock portfolios, answer telephones and turn on morning coffee.

At this point I have to split readers into two camps: those who already have a computer and want to make use of it to control a videodisc player and those who are starting from scratch. For the computer-experienced group, parts of this chapter might be a bit redundant because I will address readers who are plunging into both the videodisc and computer worlds. I will focus on computer-specific authoring and control programs in Chapter 3.

CHOOSING A SYSTEM

As you have probably ascertained by now, buying a computer requires a bit more thought than purchasing a disc player. The classic method for depleting your

25

bank account by at least a couple of thousand dollars is to find the software that seems to solve your problems, and then find a computer that runs that software. This is an ideal and a bit of advice not usually taken seriously. So many times, I've been asked by friends and fellow workers: "I've just bought a ByteBelcher 2000. What's some good software to buy?" Most successful purchases involve considerable research on the part of the buyer concerning the software-hardware interface.

The best way to go about purchasing a system is to analyze your required work load, head for the computer store, explain your needs to the salesperson and sit through some software demonstrations. Recommendations from knowledgeable friends, reviews in computer magazines and hands-on tests at the stores will guide you toward a purchase that will do the job. Not so with videodisc software-hardware packages.

The market for these programs is still small compared to the vast amount of standard computer applications products. For the most part, videodisc control systems are available through the mail, from a salesperson or from product demonstration seminars. Comparisons are difficult to make and, as with better software, the cost of these systems is considerable.

The job is easier if you already have a computer and are committed to its use because you can simply cast about for programs written specifically for your machine.

If you are looking for a complete turnkey system, perhaps some of the information sources listed in Appendix D at the back of this book will be of service.

Consider, for instance, computers that are designed specifically for interfacing with specialty peripherals. The NCR Corp. of Dayton, OH, has introduced the InteracTV, which interfaces an NCR personal computer with a Pioneer LD-V6000 videodisc player to combine text and video images. You can even get a touch screen to accompany the package.

Sony developed the first true turnkey system—the SMC-70 coupled to the LDP 1000 disc player. The latest Sony View System combines the Sony SMC 3000 computer and one of four variations of the Sony LDP 2000 disc player. We will look at the Sony system in some detail in Chapter 8. Now we will look at the IBM PC and the Apple because most videodisc software is written for one or the other.

THE APPLE

"Apple" sounds cute and lovable and was a trendy phenomenon when it was introduced to the mass marketplace. Once upon a time, Apple was king.

The basic machine surfaced in 1976, won instant success as a kit, then achieved immortality in its refined configuration, the Apple II+. What gave the design its staying power while other efforts floundered in red ink was its operating system that made writing software an easy proposition. The Apple also used a derivation of Microsoft BASIC called "AppleSoft BASIC," which was easy to learn.

A feature that greatly increased the Apple's utility was its top that was made to come off, revealing a row of slots along the back of the main circuit board. These slots allowed other circuit boards to be inserted, expanding the capabilities of the basic machine. In effect, the Apple was designed to be redesigned to fit the needs of the user.

This notion was not new. Other computers incorporated an "S-100" bus that used a standardized method of connecting boards from one manufacturer into a computer made by another manufacturer. But "standardization" was not high on the list of key concepts in those early days and the S-100 bus was frequently ignored by marketing people who wanted to establish their own "standard."

It was Apple's basic but elegant architecture and user-friendliness that drew third-party developers. Apple was also generous with licensing agreements, making its new disc operating system, DOS 3.3, available to software publishers. This gregarious, "we're all in the boat together," attitude made millionaires out of almost everyone who came in contact with the golden Apple.

In the world of computers, it is software—application programs—that makes the world go 'round. Apple got off to an early start with Tandy's TRS-80 and Commodore's PET jostling for position. Apple leaped ahead, not with innovative technology but with its immense backing of thousands of applications programs ranging from the game *Beer Run* to the first Superprograms: a word processor called *AppleWriter* and something called a *spreadsheet* written by two college students. Many computer historians feel that the spreadsheet (subsequently named *VisiCalc*) was largely responsible for lifting microcomputers out of the novelty class and elevating the machines to the status of useful home and business tools.

The Apple II+, IIe and Apple-compatible Franklin are the machines I have used and understand best.

Programs are currently being written to run on Apple's newest operating system which supplants DOS 3.3. It is called ProDOS and is far more advanced than the arcane protocols of the earlier system.

The Apple is capable of more flexibility than any other computer. It has more available software and, because of those internal slots, the Apple II+ and IIe can accept circuit boards from other manufacturers. There is a good book on the market called *The Endlesss Apple*, by Charles Rubin.* In describing the endless possibilities for expanding the Apple's basic system, it necessarily discusses the 1977 technology for which users must eventually compensate.

The Apple II family has three basic shortcomings that must be kept in mind. First, by computing standards, the Apple is slow. Yes, you can soup it up with a circuit board that carries a co-microprocessor, but that costs between $300 and $400. Second, its off-the-shelf memory is limited to 256K (256,000 8-bit bytes or Kilobytes) in the IIgs (the latest Apple model). This memory can be increased by installing a memory board with up to 2 megabytes for about $200 to $400. The last problem is that Apple disk drives are short on capacity. It's true, however, that you can buy a 20-megabyte (20 million bytes) hard-disk storage unit for around $500 to $750. Software written for the Apple must cope with these basic shortcomings, and the publisher should not expect the user to shell out additional cash for hardware fixes in order to run an application program.

---

*Charles Rubin, *The Endless Apple* (Microsoft Press, 1984).

For developers of videodisc authoring and player control programs, the Apple was the only game in town until 1982. Most of the catalogs of these developers still carry a wide range of Apple interface circuit boards and systems that plug into the Apple joystick port or the cassette interface port at the back of the case. Many of the authoring languages were developed along the lines of PILOT (Programmed Inquiry, Learning Or Teaching), a language favored by computer-aided instruction (CAI) programs. PILOT gained widespread use as Apple PILOT.

Capabilities for inputting information and interacting with the Apple are numerous: you can use the keyboard, a joystick, lightpen, touch screen, special keypad or even voice recognition. All the components are currently on the market.

THE IBM PC

Despite these positive aspects of the Apple, when IBM announced its personal computer in 1982, software developers saw gold in this new machine. Today, most state-of-the-art videodisc systems are written for the IBM PC and lower-end applications or scaled down versions of the same programs reside with Apple. It was even difficult to obtain Apple-compatible programs from manufacturers for evaluation in this book because they are out of date.

In order to follow the trend and study state-of-the-art videodisc programs, I had to dig up an IBM PC. Although the PC is not a technological wizard compared with the Apple, it is faster, has a larger resident memory, has greater capacity drives, a more sophisticated operating system and the original keyboard is similar to the Apple's keyboard. It even has expansion slots with third-party manufacturers cranking out hardware accessories. As for software, however, it is far superior. Apple's inherent lack of memory imposes severe limitations on software developers. The IBM PC has no such difficulty because of its potentially larger random access memory (RAM).

The RAM is the place where all programs and data go when a program is booted and used. Using memory upgrades, 640K and more is easily achieved and no patch software is required to take advantage of the increased memory as it is required with RAM expansion boards built for the Apple. (Patch software is booted before an application program is loaded, so the program is "hooked" into the expanded memory board residing in an Apple slot.) The more RAM, the more sophisticated the programs that can be used.

INTEGRATED SOFTWARE

The IBM PC and other technologically advanced computers using the MS-DOS language can make use of new integrated software. In Chapter 1, I mentioned that a computer is capable of many tasks while a disc player only plays discs. Integrated software combines many of those tasks into a single program. One of the truly boring parts of computing is swapping discs every time you want to move from one job to another. Apple users are used to this bane.

Then along came *Lotus 1-2-3, Framework* and *Symphony,* all written in MS-DOS for the new computers using the Intel 8088, 8086, 80286 and 80386 microprocessors. These programs combined spreadsheet, word processor and a database. Not only were the three applications installed in a single system, but they also shared the same command set and could be brought up on the screen at the same time through "windows" and data could be exchanged. Only the formidable Apple Macintosh has finally accumulated a software following that compares favorably with the latest IBM compatible programs.

With these more powerful programs, much faster microprocessors and expanded memory, the IBM PC and its "compatibles" such as Compaq, Zenith, Tandy, Leading Edge and the AT&T series became the industry standard. This fact was not lost on our videodisc system developers.

MSX COMPUTERS

MSX (Microsoft Extended Basic) Computers are designed to be peripherals as well as controllers and they promise some surprises even if their basic design was built on technology that had become somewhat frayed at the cuffs.

Regardless of manufacturer, the MSX machines resemble each other. Hitachi, JVC, Sony, Casio, Pioneer and Yamaha form an integrated family of inexpensive computers, all using the same operating system written by Microsoft.

The MSX language has been licensed to 18 companies (including most of Japan's leading electronics manufacturers) and is the first true standard operating system among major manufacturers. Some pundits predict a wave of these machines sweeping over the United States and dominating the market in the same way that videocassette recorders did. Many other experts are skeptical.

The first year of MSX production saw 400,000 units for sale in Japan and some 100,000 heading for Europe. As of this writing, MSX computers are still "on hold" in the U.S. market.

A few MSX machines were in use at the Consumer Electronics Show, Winter 1985, in Las Vegas as a result of Microsoft's arrangement with a dozen manufacturers. They are basic microcomputers, all scaled down in size and sporting the same diamond pattern cursor control arrows on the right side of the console. They all use the same 8-bit Zilog Z-80 chip, an upgraded version of the chip introduced in the 1970s. Since they all run the MSX operating system, a program written on a Yamaha should run on a Sony or Pioneer machine. The booth at CES 1985 was busy, but not crowded. Interest was polite, but hardly frenetic.

The problem with the MSX computer lies not with the new standard but with the apparent technology it reflects. MSX machines offer little more than the low-end crop of U.S. machines that are already being replaced by more advanced designs. The Commodore 64 is gone, both technologically and in the marketplace. The Apple IIe has a huge user base, but it represents 1977 technology and has been discontinued. The MSX machines are in the same league as these one-time world beaters. Today, IBM and the more advanced Apple Macintosh computers account for the lion's share of the over-$1000 market. The Atari 500ST is neck in neck with the graphics-oriented Commodore Amiga with its small but faithful following.

The market for the MSX computer seems to lie in the home computer arena. The English MSX licensees even advertise their products as "home computers from household names." Each machine, though it uses the same operating system, offers something a bit different. Yamaha ties its machine into its great success with music keyboards and their inevitable computer control. Pioneer announced an MSX machine on April 20, 1984, with a built-in videodisc interface for less than $400.

Price is another market lever. Kay Nishi of Microsoft estimated that the machines would sell in the United States for between $700 and $800, with discounts chopping about $200 off that amount. This would have placed the MSX computers in a field dotted with graves from the earlier U.S. home computer war that left Americans with a "computer in every closet."

Part of the marketing strategy for the MSX computers will certainly include their use in conjunction with videodisc and digital audiodisc players. The Japanese have never entered a U.S. market unless they were fairly certain they could dominate it and their track record demonstrates that—e.g., the transistor portable radio, the electronic wristwatch, the pocket calculator, color television, 35mm camera and the videocassette recorder. MSX licensees talked about interconnected peripherals that included the computer as part of the system when it's not being used to balance the checkbook. A videodisc player combined with the new digital TV and an audiodisc system could form quite a sophisticated package with the addition of an MSX type of computer.

Today, however, in the consumer market most microprocessors are being added to the consumer products themselves rather than as part of stand-alone keyboard computers. Note the "smart" TV sets, VCRs, electronic pianos and telephones.

In the search for the videodisc-microcomputer connection, Digital Research—the company that gave us CP/M from the pen of Gary Kildall—is the only U.S. manufacturer working on an MSX Level 3 interface. In the September 1984 issue of *Popular Computing,* Kildall said that you have to think of the videodisc-computer market in a different way, that ". . .the peripheral is not the videodisc. . .the peripheral is the computer." Even Sony has developed its still-to-be-announced MSX "Hit Bit" computer as an inexpensive videodisc controller in the consumer marketplace.

It would seem that the developer of an MSX-type computer that relies on the American consumer to use it as a controller for new, sophisticated videodisc machines should keep an eye on the new video and audio combination discs appearing on the market and the sophisticated players designed to take advantage of these new technologies. Machines such as the CLD 1030 from Pioneer that combines videodisc, CD audiodisc and CD-V audio-video playback could kill the home computer as peripheral controller market completely.

MINICOMPUTERS

On the other end of the cost spectrum, if your company is really serious—because only a company or university can afford the tab—then the minicomputer system might fill the bill.

A system produced by Terak offers a PDP 11-23 microprocessor along with a standard 512K bytes of RAM using a 10-megabyte hard disk for storage. The package comes connected to a videodisc player. A graphics package offers 25 colors on a screen boasting 640 x 480 pixel resolution and has the ability to create a video overlay. The lot goes for upwards of $20,000 and more.

Another possibility, using the same PDP 11-23 processor, is the DEC (Digital Equipment Corp.) Pro-350, introduced in May 1983. The system uses DEC's IVIS (Interactive Video Information System) language which manipulates signals from a Sony disc player. The video screen overlay resolves at 960 pixels by 240 lines and costs about $18,600 with a membrane touch panel for the monitor.

If your pocketbook is already deflated, then to tell you that a larger computer—a VAX system—is required to write software for IVIS is probably no help at all. The VAX can be had for as little as $10,000 (the MicroVAX) or as much as $60,000 for the room-size VAX 8600. This huge front-end investment, according to one company that made the plunge, was more than amortized once interactive programs were produced on a large scale. Using IVIS and the VAX authoring system, interactive training programs can be delivered for systems as diverse as Sony SMC-3000, Apple IIgs, IBM PC and even a Commodore 128 connected to a dusted off RCA CED disc player.

COMPUTER GRAPHICS

One of two possible outcomes will occur when a computer controls the disc player. The computer questions, answers, menu choices, etc., can appear on a separate computer screen monitor, or the computer text/graphics can be overlayed onto the same monitor screen. This overlaying is not an easy effect to come by, but the following section discusses how it works.

If both signals being sent to the same monitor are NTSC (National Television Standards Committee), or video signals, they have to be synchronized (i.e., locked together vertically and horizontally and chroma matched). Without synchronization, the picture will flicker, scroll, flop over and otherwise act up.

There are two ways to achieve this sync: either the disc player can be locked to the computer overlay system or the computer can be locked to the video player. Sync locking often uses what is called a timing loop to generate horizontal and vertical sync pulses, endlessly matching these vital alignments as the pictures are scanned onto the screen.

Locking up the chroma is another matter. One problem is pixel size. Since the size of the individual computer pixels (dabs of light on the screen that combine to create text or graphics) may be almost the same size as the video chroma pixels, it is important that the timing loop, or "clock," that controls the pixel scan not drift in relation to the video chroma signal or the colors could fluctuate. Another difficulty comes in with the relationship of the chroma signal to the black-and-white signal. This chroma keeps jumping in and out of synchronization as a matter of normal operation. Its phase changes don't matter when video motion segments are being played, because it takes about four fields to compensate for the shift. This compensation takes place before we notice anything when overlaying motion segments. The problem occurs with laser discs when a still frame is a true, single frame made up of two fields in sync. The chroma is then in and out of phase as the disc whirls at 1800 RPM.

It is the job of the computer interface to control the overlay when the disc player is playing either a still frame or showing video motion.

The most convenient way out of the dilemma is to convert the computer signal into a video-compatible signal so that the final screen image is actually one composite picture, properly interlaced. This text-video image can then be recorded on tape, or printed out with a video printer such as the Mitsubishi P50U.

This printer uses thermal paper in rolls about four inches wide. It simply plugs into the line between the videodisc player and the monitor and can be used to capture video images. The Mitsubishi is very easy to load and operate with a single-button remote control. When time code or frame numbers are appearing on the screen, the printer is capable of capturing a single frame and producing a black-and-white image.

This system is ideal for creating a storyboard of single frames while designing the interactive program. The images are very stable over a period of months, though they are not as crisp as the high-resolution screen pictures. Its ability to capture combined computer and video screens provides a complete record of the program for easy reference.

The composite video screen solution, however, does not take advantage of the exceptional high resolution of computer graphics (or RGB—Red, Green, Blue) compared to the less sharp NTSC images that display a limited range of detail. It is possible to convert the NTSC signal to RGB, which will give exceptional images, but that requires an expensive monitor or circuit board.

Systems discussed later in the book adopt a variety of solutions. For now, it is only important that you know the basics.

3 Authoring Languages

One aspect of computers has to be covered in broad-brush terms before we press on. The computer makes use of a binary language to operate all the tiny switches in the microprocessor and these switches allow the computer to deal with the inputs. In order to write a computer program that will control both the computer and a videodisc player, a language must be used that both the user and the machine understand. A number of these languages exist.

A vital part of any language is its ability to be used with a minimum of fuss in order to accomplish the broadest range of tasks. Developers of videodisc authoring languages have to understand that their intended users are, for the most part, not deeply committed to computing per se, but simply want to produce a program that will serve a practical purpose. Most general-purpose programming languages are very complex compared with the relatively few commands needed to create an interactive video experience. Therefore, many video control programs are written in "authoring" languages.

There is a vast gulf between what some software developers consider user-friendly and what computer-curious people consider gibberish. Authoring a computer program requires a tool kit that replaces the binary syntax with a simple, menu-prompted command set.

Some developers feel they can deduce the needs of would-be program authors and systematize the command set almost completely, limiting the options for truly flexible interactivity. We are actually still in an inductive phase in this new medium, groping about and trying to discover what this computer/videodisc connection can accomplish. There are literally dozens of these authoring programs floating about. Every company that comes up with a computer interface system to control a videodisc player takes a shot at creating its own computer program to perform the tasks required for true interactivity.

After considerable searching I found two systems that define the types of authoring programs. The first system aligns three categories:

1. The Authoring Language. This program is totally prompt-driven. It asks the user at every step, "Now, what do you want to do next?"
2. The Authoring System. This is a structure built on rules—lots of rules that replace the prompts with alphanumerics and symbols to accomplish computer-player tasks. Usually, this type of language is more flexible than the Authoring Language.
3. The Authoring Aid. This is a tool kit that provides the already knowledgeable programmer with algorithms that apply specifically to interactive video. The language is usually a thinly modified general-purpose BASIC or PASCAL. This is the most flexible, but also the most difficult to learn.

The other system is a tad more sophisticated and covers the ground a bit more thoroughly. The ground is broken into four groups.

Group 1 is concerned with simply controlling the hardware—the disc player—and is written in either machine code or a general-purpose language such as BASIC or PASCAL. It is useful for minimum applications or where speed of access is important.

Group 2 languages are full-fledged interactive video developers, but they require programming skills and a grounding in interactive program design. Most early languages, such as Tutor, fit into this category. Tutor was designed for the Control Data Corp.'s CDC 6000 mainframe computers back in the 1970s. Digital Equipment Co. offers the DAL language which runs on the VAX system and is a direct descendant of the original Tutor. A more widely known Group 2 language is PILOT in its many microcomputer variations. We'll look at PILOT in detail later in the chapter.

Group 3 offers modules or templates that the author can use to assemble an interactive program. In this group the author is confined to the design of the modules, which in turn dictates the design of the program. Some find this restricting.

Group 4 does not restrict the author to any specific instruction sets and is the most sophisticated of the lot. Its elements are broken down into sets of tools that are useful to the novice author as well as to the experienced programmer, who can create custom instruction components. An example of a Group 4 language is the CDS (Courseware Design System) designed by Dr. James Schuyler. It required two years to complete between 1973 and 1975. CDS offers a Group 2 language similar to Tutor, but lessons created in CDS can generate other lessons. Microcomputer versions of CDS have been written by Electronic Informations Systems, Inc. They take advantage of CDS's capability to work on a number of different machines. Systems have been created for Sony, IBM, Zenith and Apple's Macintosh.

PILOT

PILOT is an early language that has survived and even proliferated among interactive system developers. Its most popular version is Apple Super-PILOT which has been a mainstay with systems using the Apple as the controlling microcomputer.

As mentioned in Chapter 2, PILOT stands for Programmed Inquiry, Learning Or Teaching. It was written by John Starkweather in 1968 at the University of

California Medical Center in San Francisco. It is an easy language to learn because it is not mathematically based, as are general-purpose languages such as BASIC, FORTRAN and PASCAL.

In its latest form, PILOT is revised and more powerful than its original format. It is available for such diverse machines as the Apple, IBM PC, Atari and the Morrow Designs Micro Decision. The language was developed before interactive video came on the scene, so its primary use was for CAI (Computer-Aided Instruction) projects. Pupils are plunked down in front of a computer to interact with the courseware created with a computer language. Originally, the screens were only capable of text manipulation, but later versions of PILOT included graphics capabilities to enhance these sessions.

PILOT is an interpreted language. This means you can write a program line such as:

10 T:This is a PILOT program

and the interpreter that was loaded into the computer when the language was booted will look at that statement character by character and interpret its instruction into binary machine code that the computer can understand and act upon. The problem with interpreted languages such as BASIC and PILOT is that they are slow by computer standards. Once that line is read, the computer looks at and interprets the next line and so on and so on.

The other type of language is compiled, which means you write out the entire program and then a compiler translates all the written instructions into binary code at one time. You then run the entire program. Since the instructions are in binary code, the computer reads them directly and program execution speed increases.

The only drawback with compiled languages occurs when there is a bug. A bug in the program's instructions (an "O" instead of a "0" for instance) makes it necessary to go back to the source code (the original program), find the bug and then go through the entire process of recompiling. Improved languages, such as Sony's BASIC/1, automatically record the instructions in compiled code, but allow the program to be tested as it is created. This helps to speed the debugging process. Both the source file—the program you typed—and the object code file—the compiled program—are stored on disk. The object file is the one used by the computer's operating system for fast execution speed.

Programming Shorthand

There are many versions of PILOT and those used by interactive video developers are enhanced with videodisc player commands. We can look at a generic version to see how it works.

First of all, as with BASIC, each line of program code must be preceded by a line number. This line number may be any integer between 1 and 9999. Line numbers are usually incremented in 10s so that it is possible to go back and insert something between line 50 and line 60 if necessary. The insertion becomes line 55.

There are a number of special characters that act as shorthand. Some examples follow:

```
10 T:The Pilot language is easy to use. Do you
      agree +
20 A:
30 T:A simple language is a good language! to be
      sure.
RUN
```

The result would be printed out after the RUN command is typed and the RETURN key pressed.

The Pilot language is easy to use. Do you
agree? <u>Yes</u>
A simple languge is a good language

                                         to be sure.

The "T" stands for "Print" and when combined with a colon ":" tells the computer to print out on the screen whatever follows the colon. The "+" at the end of the sentence after "agree" causes the computer to print a "?" and wait for an input on the same line as the "T" statement.

Line 20 is an "A" for Answer. Whatever you type in response to the previous question on line 10 will input into the program at this point. In this case, the input is either <u>Yes</u> or <u>No.</u>

The exclamation point, "!" after "language" causes the program to skip to the next line.

Branching

The previous example demonstrated programming shorthand in PILOT. But for a program to be truly interactive there must be tools to accomplish this end. In PILOT, you can Match (M) answers typed into a list of possible answers, allowing the computer to make a decision based on what the user types in. An example would be the following:

```
10 T:How is your dog today
20 A:
30 M:Great, Fine, Dead
```

You can extend this capability by making the question ask "Yes" or "No" (Y or N).

```
10 T:How is your dog today+
20 A:
```

```
30 M:Good,Fine
40 T Y:Glad to hear that
50 T N:Sorry to hear about that.
RUN
How is you dog today?Dead
Sorry to hear about that.
```

The program had two expectations: either the user would type in one of the answers in Line 30, in which case the program would have responded: "Glad to hear that." If *neither* of these answers was typed in, the worst was feared, so the program answered, "Sorry to hear about that." This simple example could be expanded to include a list of possible correct answers and if none was chosen the program could "jump" to another set of questions that might go over the ground in a different manner to elicit the proper response when the former question is proposed once again. This is called branching and, in PILOT, the "J:" statement handles this chore.

As in most programs, statements are acted upon in line number order. Some times you will want to jump to another part of the program, as in the example above. You can use the Y and N as conditions for a jump in the same manner as with the M: (Match) statement.

Example:

```
10 T:In the night something climbed in my window
    |
    |
20 T:and drank my blood. What is it+
    |
    |
30 A:
    |
    |
40 M:Vampire
    |
    |
50 J Y:End (Guess right and the program ends)
    |
    |
60 T N:Sorry, that's not quite correct
    |
    |
70 T:It has wings and comes from Transylvania
    |
    |
80 A:
    |
    |
```

```
  90 M:Vampire
     |
     |
 100 T Y: How brilliant you are
     |
     |
 110 T N:Sorry, still not right. Another clue
     |
     |
200 *End T:How about that! You got it!
```

You can see that the program jumps ahead to the end at line 200 if the answer is correct at line 50. If not, another clue is given. Of course, there could be more clues, but we're discussing branching, not vampires. Branching is, to a large degree, what interactive programming is all about. The user is offered a choice of answers and the direction the program takes from there depends on the user's choice. Line 110 could have read:

```
110 J N: Sorry go to remedial vampirism



500 *End T:Welcome to remedial vampirism.
510 T: Open your book to page 27.
```

In this case the label "*End" stops the jump routine. The label could have been more descriptive:

```
110 J N:*Vampire


500 *Vampire T:Sorry. You lose your soul.
```

Remarks

Programmers like to leave themselves and their progeny small notes within programs to leave a hint about what they had in mind as they were coding instructions. These are called remarks (R). In PILOT, they are inserted as line numbers but are not printed out when the program is run:

```
10 R:Branch down to goat wrestling on line 3000
```

Repetition

Another forte of computers is their ability to repeat a statement tirelessly.

```
10 *Begin T:I am brilliant
20            T:I am wise
30          J:*Begin
```

This program will endlessly repeat:

> I am brilliant
> I am wise
> I am brilliant
> I am wise

These repetitions of the "loop" can be controlled by rewriting line number 30 to include a quantifying number and a "!" signifying "jump and repeat five times":

```
20            T: I am wise
30            J!5:
```

Personalization

As authoring languages go, PILOT provides an easy enough syntax and was obviously designed to implement training programs with a minimum of difficulty. It even allows personalization through the use of a "variable," in this case the name of the user. The name will be printed on the screen wherever the variable is inserted in the program.

```
10 T:Please type your name.
20 A:$Name
30 T:Raise your left foot six inches $Name
RUN
Please type your name.
Fred
Raise your left foot six inches Fred
```

In its generic form, PILOT is excellent for instructor-written CAI functions, but videodisc player commands have to be added to its syntax for our purposes. Its structure requires some knowledge of programming style and has definite limitations for wide-ranging applications.

Apple SuperPILOT

Apple SuperPILOT gives PILOT the ability to control a videodisc player. When PILOT emerged as an educational tool, it was a text-only system and somewhat tedious for use with children. Apple PILOT was created in 1980 to add graphics, high-resolution character sets and sound effects to remedial PILOT. At about the same time, Great Minds were turning toward the control of videodisc players. Apple noted this and sent Apple PILOT back to the drawing board. What emerged was Apple SuperPILOT. This language has endured as a staple of many interface manufacturers.

Essentially, SuperPILOT is a subset of Apple PILOT, which means a lesson written for straight PILOT will also run on a machine using SuperPILOT. What can it do? It can lay down text on a screen in 27 different colors and in single- or double-sized fonts; it can animate blocks of custom-drawn characters at 30 frames per second, and it can run a printer and help with student recordkeeping, all at a considerably faster clip than Apple PILOT.

But the most important aspect of SuperPILOT is the "V:" command. Remember "T:" meant print out on the screen whatever followed? The V: command controls a videodisc player. The command format is similar to the SONY command set (discussed in the next section) but is somewhat less intimidating. For example, "V:INIT" initializes the disc player so it is ready to receive further instructions.

If you want to find a particular frame of video, you type: "V:FIND (frame number); VIDEO." If you want to see an entire segment, then type: "V:PLAY(beginning frame number, ending frame number)."

PASCAL 1.1 is the native tongue of SuperPILOT and each lesson floppy disk contains a library of PASCAL commands in the form of subroutines (small programs within a larger program). These subroutines are used by the SuperPILOT interpreter as it runs the lessons you create. An example of these subroutines is in Library Unit 8, named "VCONTROL." Inside that unit is one subroutine called "VCOMMAND." Whenever you type a "V:" command in the program, this VCOMMAND is called up and your instructions are conveyed to the videodisc player.

SuperPILOT executes its instructions rapidly because PASCAL is a tightly structured language and its conventions are strictly prescribed. A PASCAL programmer must be a disciplined human being.

SONY BASIC/1

In the meantime, a less limiting, far more ambitious authoring system is available from Sony. BASIC/1 is designed for general-purpose programming, but Sony has created an extensive set of commands that allow complete manipulation of a videodisc player or a number of videodisc players.

To attempt to give even a broad-brush look at BASIC/1 would take far too long for our purposes here. BASIC programming tutorials abound on library shelves, so I will confine this discussion to the commands and those bits of BASIC that have an impact on our videodisc controller needs.

The extended command set requires concentration, but its structure is very elegant and simple to understand after a bit of homework.

To boot the utility, the BASIC/1 disk is started up in disk drive **A** of the Sony SMC-2000-3000 computer. Once up and running, a cursor prompt appears on the screen, showing us the BASIC/1 language is running. The next step is to load our link between BASIC and videodisc player. Type "LDP, PAC" and you are ready to begin.

The Sony View System

Sony's videodisc control language was developed from their original Disc BASIC written for the SMC-70 computer, which ran the discontinued CP/M (Control Program for Microcomputers) operating system. BASIC/1 operates in the IBM compatible MS-DOS, or PC-DOS environment. The latest version is designed to run on the Sony "View" system, which incorporates the Sony LDP-2000, or LDP-1500 videodisc players with the Sony SMC-3000V microcomputer. This system will be discussed later.

BASIC/1 is a very sophisticated language that takes advantage of the View system's capabilities, which include overlaying graphics, touch-pen and mouse control as well as sound generation. Turnkey systems, such as the Sony and IBM Info-Window, eliminate the interface problems that can occur between equipment of different manufacturers. This congenial handshaking also permits the creation of very powerful, proprietary languages and operating systems.

To control a disc player, Sony has added a set of commands, which are interwoven into BASIC programming conventions:

- LDPCONT
- LDPCSTART
- LDPEND
- LDPINIT
- LDPPAUSE
- LDPSRCH
- LDPSTAT
- LDPSTOP
- LDPCSRCH
- LDPDISCID
- LDPFRM
- LDPOUT
- LDPRESET
- LDPSTART
- LDPSTEP
- LDPWAIT

These commands are "called" by prefacing each with a "_" symbol, but cannot be implemented using BASIC's standard "CALL" command. The proper line number convention would be:

```
10 _LDPINIT: REM  (Initialize the videodisc)
20 INPUT X: REM (Variable "X" equals an input number)
30 For I = 1 to X: REM (Perform the following X times)
40 _LDPCSRCH(10): REM (Search to Chapter 10)
50 _LDPSTART(10, , , ): REM (Play back Chapter 10)
60 _LDPPAUSE: REM (Pause the videodisc for more input)
70 Next I: REM (Jump back to Line 30)
```

The following is a brief description of each command's job:

1. _LDPINIT—This command should precede all others when a videodisc player is to be controlled. It initializes both the control command set and the player.

2. _LDPCONT—Resumes suspended playback of a paused videodisc.

3. _LDPCSRCH—Searches for the beginning of a specified chapter, for example, (_LDPCSRCH(10).

4. _LDPSTART—Starts playback of a specified chapter, allowing control of how the chapter is played. _LDPSTART(10,3,2,5) means play back Chapter 10 as specified by how fast it should be stepped through (3), which is half normal speed (2), and play it over again five (5) times.

5. _LDPDISCID—Obtains the videodisc's Identification code and assigns it to a string variable. This is handy when a program requires more than one videodisc. The ID code is assigned by the producer at the time of disc mastering. The code can include information pertinent to the program, such as the first frame of video subject matter, the release number, or audio channel language code.

   For example, SOUTER DISC—10:ES:25:500:47323 means: the disc title is SOUTER DISC-10, the language channels have English on 1 and Swahili on 2, the release number is 25 and the program begins at frame 500 and ends at frame 47323. The programming convention could be:

```
10 _LDPDISCID(ID$):   REM "ID$" (The information code up to 40
      ASCII characters long.)
20 PRINT ID$
```

6. _LDPEND—Holds the execution of a program until a video disc segment finishes playing.

```
10 _LDPCSTART(10,20):  REM (Start playing Chapter 10 and end at
      the beginning of Chapter 20.)
20 _LDPEND
30 PRINT "PLAY END": REM (If line 20 was omitted, the words
      "PLAY END" would appear as soon as Chapter 10 began playing.)
```

7. _LDPFRM—Receives the current frame number and assigns it to a floating point variable.

    > 10 _LDFRM(F!): REM ("F!" can be any number between 1 and 54000.)
    > 20 PRINT F!: REM (If 5000 is displayed on the screen, then frame 5000 is being played back.)

8. _LDPOUT—Sends a one byte command code to the videodisc player.

    > 10 _LDPOUT(&H46): REM (Turn audio channel 1 on.)

9. _LDPPAUSE—Suspends playback of the videodisc and obtains a still picutre of that frame. _LDPCONT resumes play.

10. _LDPRESET—Reset the error status. If there is an error in almost any LDP command, the LDPRESET or LDPINIT command must be executed. This command is used with the BASIC programming command, ON ERROR GOTO.

11. _LDPSRCH—Searches for a specific frame number.

    > 10 _LDPSRCH(5005): REM (Search for frame number 5005.)

12. _LDPSTART—Starts playback at a frame number.

    > 10 _LDPSTART(5005, , ,): REM (Normal speed playback.)

13. _LDPSTAT—Reads the status of the videodisc player. The player's status is read by means of an array variable, which is declared at the beginning of the program in a DIM statement, such as (DIM S%(4)). This statement allows the status of the player to be ascertained at any time using the _LDPSTAT (S%( )) command. This status can show things as: Motor on or off, Program Display mode, Search and Repeat Mode, or No Disc.

14. _LDPSTEP—Moves the the next (_LDPSTEP(0)), or previous (_LDPSTEP (1)) frame.

15. _LDPSTOP—Stops playing the videodisc. _LDPCONT resumes play.

16. _LDPWAIT—Holds the execution of a statement until the player finds a particular frame number, then transfers control back to the next program statement.

    > 10 _LDPWAIT(2346): REM (Program hold until frame 2346 is found, then let the BASIC program continue.)

Another powerful set of commands allows still frame audio to be played if a still frame audio board is installed in the SMC-3000V computer. These use an "_SFA" prefix. Graphic overlays and artwork can be called up using the "_G" prefix commands and Sony's intelligent color monitor, which comes with the View system.

The Sony BASIC/1 language is a "living" language that is constantly updated with new versions to meet new challenges. It is extremely flexible and is part of a family of Sony control languages.

THE IBM SYSTEM

Another example of a disc control language system is provided by IBM along with their InfoWindow work station. IBM offers the Learning System/1 (LS/1), which claims the flexibility of an authoring language while providing ease of use commonly associated with an authoring system—usually a menu-driven shell program where the actual language is hidden from the user. Also available are the Video/ Passage Author and Presenter programs. The Author program makes use of a "spreadsheet" approach to aid in positioning the interactive events: visuals, text, sound, graphics, and so on. The Presenter allows complex branching, which makes the events truly interactive with the user.

A third IBM programming tool is the InfoWindow PILOT Authoring and Presentation system. This system makes use of the PILOT language, discussed in Chapter 2, to create interactive programs that take advantage of InfoWindow features and then record student responses and progress. It is a teacher-oriented programming system, used where PILOT is already a popular computer-aided instruction tool.

SPECIAL COMMAND SETS

The use of special command sets is common throughout the videodisc interface industry. These interface *black boxes* are necessary to connect computers to players of different manufacture.

For instance, I recently completed a video wall program for Nissan Motors that involved live talent, a videodisc program spread across eight screens, fog machines, aircraft landing lights, cassette music playback and 78,000 watts of sound power to physically move the air within the display area. All these elements were driven by a single Apple IIe computer running a BASIC program. Four videodisc players were part of this rich mix. They were controlled with four Apple Super Serial cards running through four System Impact VID-232 interface boxes. In order to access the players, the VID-232 boxes required proprietary commands such as "INIT," "FIND" and "PLAY" be embedded in the standard BASIC program. A homemade card packed with inexpensive Radio Shack relays controlled the other show elements. The program was initiated by pushing a button.

An interesting point here—BASIC works fine as a controlling language for players, even though faster executing compiled and direct machine languages can be used. The microprocessors used in videodisc players require more time to read instructions sent by the host computer than it takes for slow, old interpreted BASIC to send them.

CONCLUSION

Usually, the creators of interactive videodiscs do not require the complexity of general purpose programming languages, and since relatively few commands are needed to create an interactive video experience, many video control programs are written in authoring languages.

In as much as the medium is still relatively new, every company that comes up with a computer interface system to control a videodisc player is trying to discover just what the computer/videodisc connection can accomplish. Authoring language systems run the gamut from relative simplicity to great sophistication. Some developers of videodisc authoring languages feel that they can systematize the command set almost completely. This approach, however, limits the options for truly flexible interactivity.

The most sophisticated system, on the other hand, does not restrict the author to any specific command sets. Its elements are broken down into sets of tools that enable the novice author or the experienced programmer to create custom instruction components.

There are literally dozens of these authoring programs available. The programs reviewed in this chapter represent only a small sampling of the market, therefore the user has many possible options.

4 Videodisc Control

One word that has cropped up continually during our discussion of languages is "control." What is meant by "controlling" a disc player? We have seen the commands in some brief examples, but what are we trying to accomplish?

As you might have noticed, having come this far, we are working our way south. We began in the chilly climates of yesteryear: history, dates and accomplishments of those gone before. Next came a brief examination of the brains of the operation— the calculating circuits of computers and who's machine does what. Eventually, we gave the machine a voice through programming languages so it could speak to the somewhat less bright videodisc player. The next step toward the sunny, promised land of real interactivity between human and machine is to define what that computer voice can expect from its monomaniacal partner.

In 1978, the Nebraska Videodisc Design/Production Group was funded with grants from the Corporation for Public Broadcasting. By 1982, the Group had become financially independent through the design and production of videodiscs for clients in the private sector, educational institutions and government agencies. The Nebraska Group has probably been responsible for more dissemination of information on all aspects of videodisc technology than any other organization.

It was also responsible for identifying what have come to be accepted as the three levels of videodisc interactivity. Many authorities in the field feel the Nebraska levels have been outdated by advances in videodisc player technology. This outdating has occurred in a gray area between Level 2 and Level 3. There are also new theories concerning these "levels," such as Gayeski and Williams' six levels of interactivity, or Rhodes' continuum of interactivity from proactive to reactive.

After a while, you can end up asking how many angels can dance on the head of a pin. As with any educational theory, a new doctorate thesis hits the street every day. For our purposes, we'll examine the roots of the theory—three levels—and you can select the spin-off of your choice later.

LEVEL 1

This is the level most familiar to the videodisc-buying public. The disc player allows users to scan forward or backward and utilize dual audio channels (example:

both English and foreign-language sound tracks). There is limited random access, very limited memory and no ability to be programmed either with its own remote control or through a computer.

Exact track locations to begin a sequence (chapter search) is possible. These time-based codes allow the user to find a chapter, such as the chapter in a book. Instead of searching to frame 36,247, the user presses "Search Chapter 4" and, every time, the player will go to that track. These codes, however, are in the vertical interval (black bar between frames), not in data dumps as they are in the next category.

LEVEL 2

The Level 2 disc player is a true industrial-educational model that possesses all the attributes of the Level 1 player and adds some enhancements. This machine has full random frame access and a built-in programmable memory. The discs used in Level 2 can be encoded with short computer programs that are "dumped ' into the player's very basic microprocessor. The capacity of these chips ranges from about 1K, or 1024 bytes, to 5K, 7K and upward for new Sony and Pioneer models.

A program on the disc can direct the user to make, for example, three choices from a menu of frame and chapter numbers (possibly in the form of 1,2,3, or a,b,c). The user can punch in the selection on a keypad and the microprocessor searches out the appropriate frame. Today, very complicated programs are possible with modern keypad encoders.

LEVEL 3

This is the part that most interests us. Level 3 players have all the features of Levels 1 and 2 players, but are unique because they offer control by external microprocessors. Large computer programs can be used to combine text and graphics with video pictures in still-frame or real-time motion. While Level 2 offers random access of frames and text, the videodisc programs are immutable except to create a new disc with updated information. Using Level 3, the computer text residing on an easily modified magnetic disc can be updated without changing the images on the videodisc. This capability allows great flexibility and economy.

Another attribute of Level 3 is the ability to store the scores, the responses and the paths by which users selected their responses. This allows polling of the users' data and better information on how the interactive program was actually used.

Technology has added yet another possibility to microcomputer control in Level 3. The perfected ability to store huge amounts of digital data on an optical disc format now allows the controlling microcomputer to receive its instructions directly from the videodisc. The Sony View 2000 and 3000 systems permit this type of data transfer. Compared to the small digital "dumps" (1–10 thousand bytes) fed to the videodisc player's microprocessor in Level 2, or the floppy disk's maximum capacity (1–3 million bytes), the amount of digital data stored on a disc's audio track can be measured in billions of bytes. Its main disadvantage is that the data is immutable.

Once laid down on the disc, it cannot be changed by the programmer or user. Floppy disk data is readily changeable, either directly through the computer's keyboard or by way of a telephone modem file transfer.

As optically written and read data become more flexible and accessible, the possibilities of interactive disc control will become even more challenging.

CLV AND CAV DISCS

The following section on discs discusses the two types that will play on laser players: CLV and CAV discs.

CLV (constant linear velocity) discs play for 60 minutes to a side and the disc speed varies from 1800 RPM at the inside radius to about 600 RPM at the disc's edge. The packing of frames on the tracks to conserve space makes it impossible for the laser beam to isolate a single frame of video on a single track. Therefore, the CLV disc cannot be played with freeze-frame, step motion, slow motion, frame searches or picture stops. These discs are used primarily for movies and constant play programs. A basic Level 1 player will run these discs from front to back with stereo sound and images superior to VHS or Beta videotape.

Each track on a CAV (constant angular velocity) disc contains two video fields that make up one video frame. The disc's speed is constant at 1800 RPM and playing time is restricted to 30 minutes. Looking at the surface of the disc, you can see two wedges of lines cutting across the surface. These are the vertical intervals where the scan returns to the top of the screen to trace down the second field of video. When the laser finishes reading one track from the inside, it moves out on the spiral to read the next track in typical linear fashion.

This laser can then be commanded to proceed to the next field, or jump inward or outward to any other frame. That is the big difference between CAV and CLV discs. Each of the 54,000 frames on a CAV disc can be considered a separate entity. Since the image is read with a beam of light, playing time for a single frame can be indefinite without damage to the record.

Step frame is a controlled version of freeze-frame with laser videodiscs. A single track of video is read until a command moves the laser to the next frame, stepping through the program one frame at a time.

Slow motion involves repeating a frame a given number of times before moving on to the next frame. This can be accessed in forward or reverse mode.

Fast motion is the opposite of slow, causing the laser to read only a single *field* at a time and speeding through a program.

Scan mode skips over several tracks at a time much like skimming through a book.

Searching can be implemented through a computer program or manual keypad. Individual frame "addresses" can be located and played in the search mode.

Laser discs can also have "chapters"—disc segments containing specific program modules. The Chapter Number Codes are placed in those vertical interval wedges between video fields.

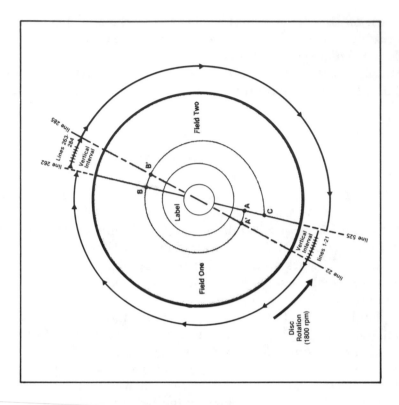

Figure 4.2:   Format of a CAV disc.

Courtesy of 3M Optical Recording

Figure 4.1:   Format of a CLV disc.

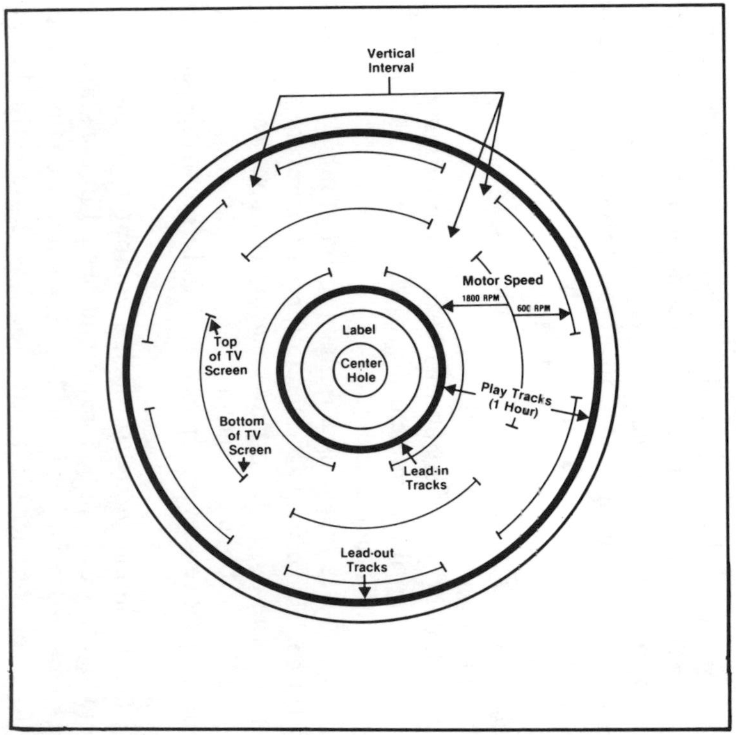

Courtesy of 3M Optical Recording

COMPUTER INPUT

We will be coming back to these levels of control and disc anatomies from time to time in our search for interactive video. A final stop at the computer input stage is necessary, though, before we look at a few examples of interactive interface systems in detail.

Control of the disc player by the computer is a matter of electronics, but control of the final product—an interactive program—requires human engineering.

For Level 1 and Level 2 players, the normal mode of inputting instructions to control the disc is through a remote keypad, or by punching buttons on the player's deck. With Level 3, the options are almost limitless.

I saw my first remote keypad for Level 2 control when I began this book. I unwrapped two Pioneer players an LD-V4200 and LD-V6000A series, and hefted their controllers. (See Figures 4.3 and 4.4.) At that point, I was ready to call my editor and tell her to forget the whole book. On the LD-V4200 pad, there was an intimidating array of function keys, alphanumeric keys, sliding analog control levers and odd, cryptic keycap symbols. This sophistication in controllers is part of that gray area between Level 2 and Level 3 I mentioned earlier. Before throwing in the sponge, I sat down with the instruction book for a few minutes and quickly discovered there was nothing to fear. I calmed down until I saw an old Sony controller for the LDP-1000A player. (See Figure 4.5.) It's either "worse," or "even more flexible," depending on your point of view. The controller for the Sony LDP-2000 player is much more user friendly.

Many program designers have overcome this problem by adding their own cover, or shell to the remote keypad. This shell covers keys that will not be used and can specify functions for the keys that remain.

Sony's LDP-1000A keypad has a flip-up plastic cover to hide some of its more esoteric functions. (When General Motors purchased interactive videodisc systems for its dealerships across the country, the company used a keypad cover to help users select the car of their dreams.)

At Level 3, the options begin to open up. Control of a Level 3 system is provided by an external computer, such as an IBM or Apple. Programming for a computer keyboard input is obviously the easy way out. Users can be asked to input letter, number, word, space bar or even cursor move responses. But, generally, they have to look away to search out the proper response key, breaking concentration with the program. If there is a choice to be made between objects on the screen (such as equipment parts, color squares, etc.), the elements must be labeled alphanumerically to match the keys.

Shuttling the cursor about with a mouse (that inverted trackball so much in vogue) or a joystick helps quite a bit. The "fire" button on the stick can be used to indicate a choice. This method is also easy to program. Its main drawback is wear and tear on the device. Always remember Murphy's Law: If Something Can Go Wrong, It Will. Joysticks and mice break.

The most user-friendly method of control is the "touch screen" and its close relative, the "light pen."

Figure 4.3:   LD-V4200 industrial laserdisc player.

Courtesy of Pioneer Video, Inc.

Figure 4.4:   LD-V6000A industrial laserdisc player.

Courtesy of Pioneer Video, Inc.

Figure 4.5: LDP-1000A videodisc player.

Courtesy of Sony Intelligent Video Systems.

Touch screens work on a number of different principles. One system uses a grid of crisscrossing, infrared light beams. When your finger breaks the grid, the exact X (horizontal) and Y (vertical) coordinates are fed to the program and the response is acted upon. Other systems simply stick a plastic encased grid over the screen surface and touching the grid compresses two layers together forming a connection at that spot. The electromagnetic drawing pads use this same method when a picture is drawn with a stylus.

While the touch screen is handy, it can also produce a grimy screen at the end of the day. But in a controlled situation such as a classroom, touch-screen inputs are a good choice and promote real interactivity with the program.

The light pen runs a close second in optimum systems. Its narrow beam of light can pinpoint small areas such as service locations on an automobile engine or areas of the brain for surgeons. The pen brings hands such as those on Julius Irving and a seven-year-old child down to a common denominator.

One problem with screen inputs is the limitation to multiple-choice and true-false questions. Some designers evoke a simplified keypad on the screen, expecting the user to input words, names, numbers, etc., by tapping the glass.

The use of icons can be helpful, but the designer must make sure the icons are universally recognizable. Icon designers at the 1976 Olympics in Montreal ran amok with sports symbols and many press folk, myself among them, rolled into venues expecting a basketball game and ended up facing two hours of Bulgarian ankle wrestling.

Voice recognition is also being tried. Again, the closed environment makes this option possible. Current voice recognition technology has difficulty with regional dialects and foreign accents.

Some program controllers are combinations of simple keypads and touch screens, enabling a variety of responses.

"True interactivity" is made possible by inserting an interface unit between the microcomputer and videodisc player. The interface is a combination of software— programming languages of the sort we have glimpsed briefly—and hardware. The systems usually come in packages and the parts cost anywhere from $50 to $20,000 and up. Versatility and features cost.

Every week a new device appears in the trade press, so any list would have to be incomplete. But the sheer diversity of types of interfacing systems is instructive if only to see what is available in the marketplace. See Appendix A at the back of this book for a list of interactive system developers.

5 Level 3 Applications

Level 3 interactive video offers the most exciting options to the user and challenges to the designer. At Level 3, you have nothing but choices: disc players, computers, interface controllers and just about any kind of user input device you can imagine. There are keyboards, touchscreens, light pens and mice, and in at least one instance cited in this chapter—a welding torch.

With options also come problems of compatibility: hardware, software, firmware, courseware. Where do you begin? The designer can turn to hundreds of current applications for inspiration. Training workstations are teaching everything from fast food service to auto mechanics. Point-of-sale kiosks vend hardware and rent cars. Museums and libraries catalog art and loan literature. You can study a Van Gogh painting at the National Art Gallery, or a beautiful Greek vase at the Getty Museum, on videodisc.

An entire book could be written about the options at Level 3. We will look at some of these and discuss how they were planned and executed. The first thing to understand is that in Level 3, the videodisc player is controlled by a microprocessor that is not part of the disc player as in Level 2. By separating the two functions, control and playback, you gain power.

THE MICROPROCESSOR

Level 3 videodisc control spans a wide area of applications. Any external microprocessor control of the videodisc player constitutes Level 3, but this does not necessarily involve a microcomputer. Computer programs can be encoded on an EPROM (Erasable Programmable Read Only Memory) chip and simply pressed into the circuit board. New programs become instantly interchangeable and are also less volatile than those required to boot up from a floppy disk and the attendant mechanical disk drive.

BASIC CONFIGURATIONS

When you consider Level 3 for your interactive applications your options are bounded by a turnkey system, such as the Sony View or IBM InfoWindow, and an alternative of three basic configurations using the interfaces previously cited.

The first and least expensive configuration, for the most part, is the two-screen configuration. This setup consists of the microcomputer, the videodisc player, an interface device that connects them and two monitors: one standard green or amber screen and one color monitor or television set. (See Figure 5.1a.)

This is a low-cost way to get into interactivity, making use of color video images to enhance CAI (computer-aided instruction) programs. Monochrome (or color if you wish) high-resolution graphics can be used in the computer program's display as well as a considerable amount of text.

Unfortunately, the trade-offs can be considerable in many applications. Since two visual displays are required, additional setup space must be found. In point-of-sale applications (which will be discussed later in this chapter), the screens are generally stacked one above the other in a kiosk. It is still necessary, even in the most successful application, to divide your attention between two focal points. Programming should try to create definite attention directors between the two screens so the user always knows where to look.

Another configuration requires the computer, a computer-controlled video switcher, a single TV set, the videodisc player and a computer-to-player interface.

You are always looking at a single screen with this system so you save the cost of having a second display. The result, however, is that you will only view either computer-generated text or graphics, or video images at one time. The switcher handles this chore, jumping back and forth between the two display modes as directed by the computer. A major drawback is the need to rely on the resolution of the video-type screen for computer text. This limits the amount and sharpness of the text displayed on the screen. (See Figure 5.1b.)

Finally, we come to the configuration that will overlay computer text upon the video images from the disc player. This requires a computer, overlay video mixer, the computer-to-player interface and a TV set or RGB (red, green and blue) monitor. (See Figure 5.1c.)

Again, we are directed to a single screen display, but with this setup the computer text and graphics can be read at the same time. This is the most expensive set-up because it requires the overlay mixer. The mixer can accomplish its task in one of three ways: it can simply overlay the graphics on the video image much like the Sony SMC 3000 system, it can convert the computer output to NTSC (National Television Standards Committee) standards, or it can convert the video images into RGB signals in the same way that the NCR InteracTV does. This conversion allows the final computer-video image to be taped or recorded on a device such as the Mitsubishi video printer because the screen image is a single, interlaced image and not a true overlay.

From the standpoint of planning and programming, overlaying computer text over video images is the most complex procedure. Trying to predict the intersection of the computer image so it blends and complements its video companion presents

Figure 5.1: Three videodisc configurations.

(a)

(b)

(c)

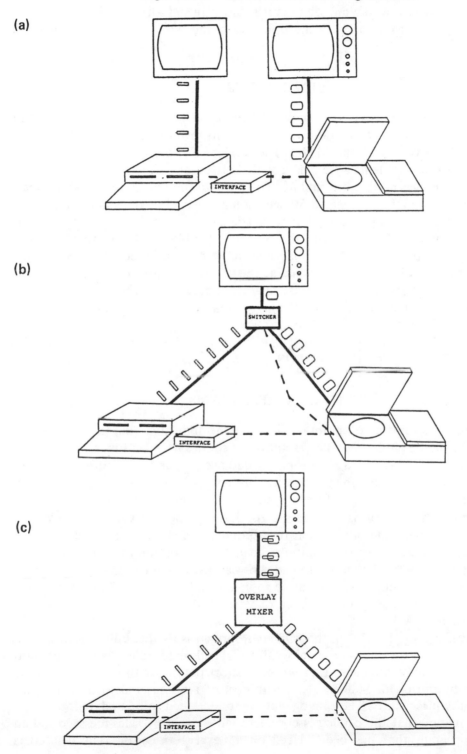

Courtesy of Pioneer Industrial Video.

problems for the designer. The variety in types of Level 3 interfaces allows many possible alternatives to overlaying. The Level 3 hardware and software evaluations presented in Chapter 8 examine both multiple-screen and overlaying systems in some detail.

VIDEO GAMES

The video game industry was considered the great hope of renewed interest in consumer videodisc use. At one point there were no fewer than 25 videodisc games on the market between 1982 and 1983. The level of technology was excellent and most of the start-up problems were licked at the outset.

The problem was a "two-bit" failure. To amortize the horrendous cost of creating the games, manufacturers charged 50 cents a play instead of the standard quarter. A player could get 10 plays on any other arcade game and only five from *Mach 3, Dragon's Lair* or *Cliff Hanger*. Also, the technical problems of resetting standard coin boxes to accept two quarters a play instead of one was a contributing factor. When the business collapsed, it fell apart at the seams. The company that produced and distributed *Cliff Hanger* ended up selling off thousands of the discs at 10 cents a piece and the used Pioneer LD-1100 players that were used to play the games went for $90 to $110 each.

However, another application that placed Level 3 videodisc systems in public areas began to flourish—point-of-sale (POS).

POINT-OF-SALE SYSTEMS

The broad definition of point-of-sale systems indicates that they require a stand-alone kiosk, or public-access workstation of some sort. (See Figure 5.2.) Inside the kiosk are one or two video monitors, a videodisc player, a microprocessor control system and some means for the user to interact with the video images. This can be a touch screen, joystick, simplified keypad or defined choice buttons.

The idea of the system is to attract possible purchasers, cause them to view a message on the screen and allow them to respond to what they see. This response can be polled and stored in the computer, can be telecommunicated to a central database for processing of a purchase order or can present locations of stores that handle the desired merchandise. The more sophisticated POS setups can even accept data from the magnetic strip on the back of your bank credit card as you slide the card through a reader.

The advantage of Level 3 in the POS environment is its flexibility. Take a catalog for instance. If you publish a catalog of merchandise complete with prices and model numbers, you can be sure that within a short period of time, models will be sold out or discontinued. Sudden changes in cost of raw materials or in supply of items will cause price shifts. The catalog will eventually need to be updated.

Now, imagine that the catalog is complete with a snappy sales message and an interactive program that has been placed on a Level 2 videodisc. The disc costs

Figure 5.2: TOUCHCOM system in a SHOP-R-AIDE-kiosk at Hess's Department Store in Allentown, PA.

Photo courtesy of Digital Techniques, Inc.

thousands of dollars to create. To update it will cost thousands more. Level 3 video-disc control, however, requires only that the information distributed by the computer microprocessor be updated. The videodisc images—the most expensive part of the process—can remain untouched. This way, prices and model numbers, addresses and available options can be readily updated at a minimum cost. Products can be deleted from the video catalog with a new computer program. The product image will still be on the disc but the viewer is never aware that it is still there.

One of the main questions a POS designer must ask is, "What type of system configuration is needed to suit this particular application?"

Budget Rent-a-Car

A company named Advance Touch asked this question as it set out to design a POS system for Budget Rent-a-Car.

Budget wanted a kiosk that would help customers decide what type of cars they wanted to rent; show them pictures of the cars; list rental prices; obtain information (driver's license number and expiration date, etc.) from the customers; read credit card information; and prepare a printed contract. In this way the customer could hand an attendant a finished document and drive away. At the end of the rental period the kiosk would also check in the cars as they were returned.

Advance Touch created the kiosk for Budget. The only problem was cost—each kiosk cost approximately $10,000. However, an increase of 2% in the amount of car insurance ordered by kiosk users due to a special promotion in the system helped pay for the units.

Two Budget kiosks are currently in operation (at the Los Angeles and San Diego airports), and an order has been placed for additional systems.

Builders Emporium

POS applications are not storming into the marketplace; they are creeping in slowly. Pilot stations are being placed in strategic locations and then monitored for results. A large POS system is used by Builders Emporium, a West Coast home center chain.

The company that put this scheme together, MarketDisc of Newport Beach, CA, did not charge Builders Emporium for the system but used an innovative marketing approach. Instead of approaching Builders Emporium with a sales pitch, MarketDisc *gave* the kiosk to the home center chain and then convinced other home improvement suppliers to advertise on this kiosk. A computer was attached to determine how many people viewed each advertiser's message and promotional material was situated close by.

The economics for networks such as that employed by Builders Emporium call for a minimum number of sponsors to be put on contract for 6 to 12 months before the project is launched.

TOUCHCOM

TOUCHCOM—a touch-screen system offered by Digital Techniques, Inc.—is an imposing slice of high tech.

The system is based around the Intel 80286 microprocessor (as used in the IBM PC/AT) with a fast clock time (interval of machine cycles). Its memory capacity is expandable to 640,000 bytes. This is more than ample for most onboard programs that remain resident in memory rather than being accessed from a storage medium such as a floppy disk. To make sure the program remains (in volatile RAM) in case of a power failure, an optional battery backup is offered.

The display is a high-resolution 10-inch screen of outstanding quality capable of 128 colors, 16 colors per RGB screen. Covering this display is the user input device, a pressure activated, membrane touch screen.

A hardware option that is becoming more evident in POS systems is included with TOUCHCOM: the telecommunications modem. A 1200-baud modem with automatic answer-dial is built in so the owner of the kiosk can call in program changes and download (remove) information concerning stored polls of user responses, user frequency data, or even system diagnostics to see that things are running as they should.

The disc player is a Pioneer LD-V6000. It is interfaced to the microprocessor in full overlay mode, allowing computer graphics over video image. As discussed earlier, this is not a true overlay since the video signals are converted from NTSC standard to RGB, which is the best way to take advantage of the high-resolution screen.

A 40-column dot matrix printer can also be built into the kiosk, which will print pages of text, forms, graphs, calculations, coupons, receipts—anything that might be offered to users as an end product of their session with the marketing message. This user interface also includes a magnetic strip reader for credit cards. User information can be stored on dual floppy disk drives, or on a hard "Winchester" disk drive with a capacity of 10 million bytes.

A workstation for writing programs is also provided. The programming or "authoring" language is called "TNT." Essentially, it is a command-driven language made up of English language macros (combinations of machine language commands) that perform the program functions. The system also includes a digitizing video camera that allows the programmer to convert slides, flat art—anything that can be photographed, into a high-resolution computer image. All things created with the authoring workstation can be downloaded to kiosks with a modem. Kiosks scattered throughout a wide geographic area can be serviced in this manner from a single marketing location.

A list of potential retail applications offered by Digital Techniques includes some interesting possibilities:

- Show all the selections in a bed and bath department without stocking excess inventory.
- Demonstrate how to use a personal computer or help select software.
- Take customers step by step through home improvement projects.
- Demonstrate industrial equipment in action.
- Help customers select eyeglass frames.
- Let customers preview movie and rock-and-roll videocassettes.

Electronistore

Many new companies are being organized to fill the growing demands for point-of-sale systems. Publishers, such as the giant R. R. Donnelley Co., are investing considerable sums in the technology. Donnelley's Electronistore System is the result of two years of development and testing. (See Figure 5.3.) It provides a two-screen kiosk, color monitor and a computer CRT, plus stereo sound, credit card reader, modem for data transmission and other options. It also offers consultation and other design services to its clients.

Figure 5.3: A point-of-sale kiosk created by Electronistore. It has two screens—one for video and one for computer data, plus a credit card reader.

Photo courtesy of Electronistore.

Ease of Use and Flexibility

The user comfort factor must be considered with any POS system. Anyone under 30 years old has grown up with computer technology and should have no trouble with this Level 3 kiosk technology. Over-30 folks—the ones with the most spendable incomes—must be catered to a bit.

The Budget Rent-a-Car system is one that is very easy to use. It has two slots: one for the credit card and one as an exit for the printed contract. The sales pitch begins when the screen is touched, but when the credit card goes into the slot, the fluff is skipped over and the microprocessor steers the program right to the heart of the matter—renting a car. The information from the card flashes on the screen and, after asking what type of car is preferred, the contract is issued from the printer in the kiosk.

Models and available cars are presented with a touchscreen interface for selection. The program walks the user through the choices, if necessary, confirming the touch-choice on the screen as verification. A touch pad is presented on the screen if a driver's license number must be entered, complete with BACKSPACE and ERASE keys for mistakes. Insurance is also handled with the touch pad.

This flexibility extends to the daily upgrading of user information. When the Budget station opens each morning, the clerks can delete or add available car models to the user's menu so the kiosk is always current.

This type of national plan for distributing Level 3 POS setups is still in its infancy. Most Level 3 systems are developed for training, or by small businesses that employ a few units. A trainer simply hooks up a videodisc player to a computer and uses the system to teach a subject. If something must be changed, the trainer can swap floppies and videodiscs and continue. This is a far different setup than a complex national network.

Touchscreens seem to have the edge in user-friendliness in the POS environment. The more expensive touchscreens are quite forgiving, allowing the user to hit reasonably close to the indicated spot. The cheaper ones require some precision in application of the finger. Keypads are susceptible to harsh treatment and have a tendency to break down.

The marketing approach in the early days of videodisc POS systems was oriented more toward sophistication and bells and whistles than to sound marketing strategies. Kiosks selling products were placed in grocery stores where shoppers were bent on getting in, buying their weekly groceries and getting out as quickly as possible. Anyone who blocked part of an aisle while trying to fiddle with an interactive kiosk ran the risk of being run down by a shopping cart.

Other experiments placed kiosks featuring discount products in shopping malls where the kiosks were direct competition to the surrounding stores. Out went the kiosks.

Many POS applications don't require huge, sophisticated kiosks. Simple, self-contained units are finding wide acceptance. Apple Computer, Inc. has touchscreen systems combined with Apple keyboards that demonstrate the ease of using Apple's line of machines. Companies such as ViMart out of Los Gatos, CA, offer one-piece

Figure 5.4: ViMart point-of-sale videodisc system in a stand-alone unit.

Photo courtesy of ViMart Corporation.

modules containing all the hardware that can sit on a shelf. (See Figure 5.4.) These systems help with stock control and its Level 3 systems (ViMart is actually a hybrid Level 2) allow for changing prices and consumer information.

EDUCATION AND TRAINING

Interactive systems put to educational and training uses tend to exist in a vacuum. After hardware is purchased for this type of application, software is often written in-house and machines are operated on campus or in a closed company network. Despite this lack of exposure, educational and training programs are capturing an increasingly larger share of the interactive videodisc market.

Educational Applications

Educators are beginning to see some advantages in the videodisc-microcomputer connection, such as the ability to teach students more cost effectively and the option to provide personal tutoring.

Cost effectiveness provided by interactive video can permeate all levels of education. Adult education for both technical and white-collar skills has absorbed some $200 billion annually and 10% to 50% of that amount is targeted for new technology that puts learning under computer control, according to WICAT Systems Inc.'s Training and Education Division in Orem, UT.

WICAT has developed systems for technical and human skills training. The latter is directed at fields in which interpersonal relationships are important, such as Sales and Counseling.

Videotape teaching systems were introduced in the 1950s, almost as soon as videotape itself was developed. A student could view a procedure, run the tape back and view it again, while answering questions on paper as the lesson proceeded. What was missing was interactivity and alternatives to single-response situations. These early systems were put down as simple "page turners."

The new, interactive videodisc programs move students through lessons at their own pace, pointing out errors and moving back to more remedial approaches to problems when necessary. Students also have the option of skipping through material they already understand.

In Level 3, the system keeps track of the student's progress throughout the program. The systems are patient, never have an off day and never lose their tempers.

Program Successes

The Omar Bradley Middle School in San Antonio, TX, conducted an experiment. Students were divided into three groups: Group One watched a linear videotape of a subject; Group Two watched the tape and discussed it with the teacher; and Group Three watched an interactive tape produced by Panasonic. The interactive system consisted of a microcomputer, word processor, keypad, a printer for statistical information print-outs and a standard videocassette recorder. The program was peppered with questions, and both rewind and stop points were electronically marked. One of the tape's audio tracks held invisible markers, while the other carried monaural sound.

Students using the interactive system registered 100% participation while the other groups were considerably less interested in the subject content. When a test was held at the conclusion of the program, the interactive Group Three scored much higher than the noninteractive groups. Its high scores were apparently due in part to the interactivity.

Designers of interactive programs claim that course completion times are reduced by 25% over conventional teaching methods and retention increased by at least 50%. Students utilizing interactive systems in a University of Florida interactivity test completed a course in 120 hours compared to 160 hours for the control group.

Keeping the costs as low as possible was the challenge for the Minnesota Educational Computing Consortium when it set out to create an interactive economics course. It chose the Apple II+ because there were so many already in the school system. A Pioneer VP1000 consumer videodisc player was also chosen. To produce the program, they planned carefully and used the video facilities of the Osseo Public School District.

The end result was a 3/4-inch videotape made up of original video, 16mm film, slides and graphic flat art. The production cost was far below the average commercial cost of $1000 to $3000 a minute. The Nebraska Videodisc Design/Production Group at the University of Nebraska was chosen for encoding the program and the facilities at 3M produced the master videodisc from the tape. The cost of post-production and mastering was about $6000.

As ingenuity dictates low-end solutions and saved dollars inspire the high-end systems, companies producing interactive Level 3 videodisc hardware and software press on. We can only hope they keep up the good work.

Training and Retraining Applications

Even though interactive videodisc training represents only a small fraction of today's training picture, it is growing at a great rate. Major companies such as IBM, AT&T, Ford, General Motors, Hewlett Packard and General Electric have established videodisc programs to aid in their heavily technological training sessions.

Training Applications in the Automotive Industry

A few years back, Ford bought 4000 Sony interactive videodisc systems and has been pleased with the results ever since. A mechanic can sit at one of the interactive training stations and be presented with a close-up view of a carburetor while using a light pen to simulate the results of a vacuum gauge and a screwdriver on an actual engine. As he works with the pen, the system's audio track furnishes actual engine sounds that rise and diminish as he works. When he believes the engine is tuned to perfect pitch, he presses the light pen against a spot on the screen and the computer sends him to a section of the program that either praises his work or brings him back to the carburetor to try again.

AMC and Toyota pioneered similar systems in their service centers.

The major cost savings lies in transportation and workers' time off the production line during extended training periods at the company training centers. If a company buys enough systems to cover its branches, the costs of program development can be amortized against traveling training expenses.

An Arc Welding Program

The ingenuity shown by some training program designers is typified by an arc welding tutorial designed by David Hon. Hon has already achieved notoriety with his design of an interactive CPR program, which uses a manikin as a control device. His arc welding program is nothing short of miraculous.

The purpose of the program is to teach students how to make a good weld. The learning experience had to parallel the actual process of welding in order to impress the basic elements of the craft on the student. It took Hon about 14 months to design and produce an interactive facsimile of the welding act.

Using an analogy of heat to light, Hon turned the video screen to a horizontal position and provided the student with a "welding torch" that produced a perfect weld across the monitor—if the student did everything right. The "torch" is actually a light pen with two adjustments on it—one representing acetylene, the other oxygen. Turning the knobs regulates the "mixture" and actually feeds data into the system. The student's choice must correspond with the program's expectations for the mixture. The student then uses the "pen-torch" to trace a joint between two steel plates pictured on the horizontal screen. If the light is held too far away or too close—according to the mixture—the weld would be faulty and the student would be branched to a remedial part of the program, which describes the proper way to overcome the particular difficulty. If everything is right, the moving torch produces a perfect weld between the two plates. The realism of the simulated welding process is uncanny.

No words appear on the screen, so all commands and lessons are verbal, allowing any two languages to be laid down on the videodisc's dual audio tracks. A Philips VP 935 disc player was used. This method of organizing the student's perceptions is called "fractionating" by Hon. With the demonstration, he combines physical and visual involvement in an accurately recreated experience.*

Interactive Applications in Business

This quiet revolution in training techniques is making inroads in other businesses as well. New employees at any of the 450 Sizzler restaurants around the country can take advantage of videodisc learning between rush hours, by wheeling out a learning module and studying procedures such as how to take an order and fill out the order slip.

J. C. Penney chose to use videodisc training for inventory control because books and printed materials on the subject were tedious. The videodisc images added an element of interest and life to the sleep-inducing material.

Book publishing companies are finding there is, if not gold, then at least a little silver in the interactive hills. Xerox Learning Systems, which has been selling educational materials to the Fortune 500 companies since 1965, packaged its first interactive course which teaches sales techniques.

Military Use of Interactive Programs

A major portion of the market at this time is made up of corporate and military high rollers, due to the expense of producing interactive programs. The average cost is around $40,000, but it can easily climb to $100,000.

---

*Taken from a teleconferencing demonstration given by the International Interactive Communications Society—Chicago Chapter, 1988. (IICS/Chicago, P.O. Box 81088, Chicago, IL 60681–0088.)

The military has turned to interactive Level 3 programs to acquaint personnel with the sophisticated tools needed by today's fighting forces. From complex radar systems to field tactics, classroom sessions eliminate building expensive simulators or wearing out even more expensive weapons systems trying to create and solve problems.

In order to come to grips with a World War II fighter plane, maintenance people had to cope with a manual some 1000 pages in length. The manual for a modern jet fighter such as the F-18 runs more than 500,000 pages. The cost of interactive learning systems is quickly amortized in such circumstances.

The U.S. Army's commitment to interactive videodisc training was emphasized in November of 1986, when the EIDS system was established. This long-range training program involved an initial contract for 1985 units and was awarded to Matrox Electronic Systems of Dorval, Canada. Each unit represented a microcomputer, videodisc player, control interface and software. It was expected that the contract could expand to include some 47,900 EIDS units over a period of four years.

A single authoring language package called EIDS-ASSIST is used for courseware development. The software was produced by the Computer Science Corporation and has been released into the public sector under the label *IV-D*.

What is newsworthy about this system is its *scale*. A major investment such as this could help establish standards that would help make videodisc applications more economically viable. In the same way that benefits dribbled down to the consumer market from NASA's forays into space, so could the EIDS program provide expertise in interactive technology for the contracting companies. These companies, in turn, would also serve the private sector. New tools, new ideas and considerable knowledge in the development of courseware could be the inheritance.

Major advances in videodisc program training would move the success rate of students even further along than the current statistics reveal.

INTERACTIVE VIDEO SOFTWARE

Level 3 applications have breathed a bit of life into the once static computer software market. The computer market is defining itself more clearly in favor of business applications. Just as the videodisc arcade games went into the dumper, computer games of the simpler "shoot-'em-up" variety are diminishing in number.

The high-end business machines IBM, Zenith, Epson, Compaq (with a long list of run-alikes) and Apple's Macintosh dominate the present market. Interfaces for Level 3 have shifted their emphasis from the low-end machines toward these heavyweight, memory-loaded machines. In a marketplace where business applications software such as Lotus 1-2-3, Framework and dBase III tend to wrap up all the buyer's needs into a few packages, Level 3 videodisc control adds a new dimension to that computer investment.

New companies are offering not only hardware but packaged applications. As with the computer market, the availability of software in this market is a driving force.

Courseware

In a survey titled *Compatibility of Interactive Videodisc Systems*, published for the Instructional Systems Association by Future Systems, Inc., the amount of courseware currently available was hinted at.*

Fifty-nine firms reported to the survey that they were videodisc courseware developers. Out of 1376 videodisc titles produced by 40 of these developers, 465 were generic titles and 911 were custom-made for clients. These titles ranged across a span of subjects, including health and medical, sales training, technical training, computer software and basic skills. Twenty-two developers claimed sales amounting to almost 100,000 copies of videodisc courseware by the end of 1986.

Most of these programs were written for the IBM Personal Computer architecture; IBM and its popular MS-DOS operating system outnumber the Apple II family of computers about four to one in the industrial market. The Apple II computers find their most fertile ground in the educational market. By 1988, however, the Apple Macintosh had made inroads into the industrial marketplace, particularly by way of the advanced Macintosh II and the *HyperCard*—a peripheral circuit board that permits users to author programs with a minimum of effort and still achieve professional results.

Integrated, turnkey systems find great favor with developers, according to the survey. Of the software in the hands of users at the time of the survey, the following picture emerged:

IBM InfoWindow	18 systems
Sony VIEW	17
Visage	13
ITS-3100	8
NCR InteracTV-2	4

Very few developers favored component systems over integrated workstations and many offered custom hardware along with their proprietary software. In fact, some 51 respondents claimed they leased or sold hardware with their courseware systems. Of all the courseware systems sold, no single authoring program, or language, dominated the rest, showing that the current programming scene is still fragmented with no clear standards at present.

CONCLUSION

At the beginning of this chapter, the breadth of options was hinted at. We have looked at a fraction of the possibilities in both hardware and software. The

---

*The survey, *Compatibility of Interactive Videodisc Systems,* is available from Future Systems, Inc., P.O. Box 26, Falls Church, VA 02246.

options offered in Level 3 are both a blessing and a curse. While the designer enjoys a variety of possible solutions to any interactive program, the medium of Level 3 interactivity is desperately in search of standards. Hopefully, major applications such as the Army's EIDS program will go a long way toward urging system developers to establish those standards. Then and only then will Level 3 really take flight and truly achieve its potential.

6 Use of Still Frames

Still frames are a major consideration in any interactive videodisc program. There are about 30 minutes of playing time for action video on a CAV disc, or 54,000 still frames. If you use only 20 minutes of that action video, you still have the remaining capacity for 18,000 still frames.

This immense capacity for still images has been taken advantage of by museums that catalog their collections, libraries, patent search services, art galleries, and slide and photo archives—any application in which image storage and retrieval must be as painless as possible. Only random access disc technologies—whether reflective, or film-based—can provide this kind of access speed and image quality, while requiring only very basic computer programming.

Still frames also play an important part in training and point-of-sale interactive programs. These frames can be text, graphics or still photography, and are often combinations of all three. The ability of a disc player to display a single disc track—one frame—indefinitely allows still frames to be part of a point-of-sale kiosk's "attract mode" (that lures a potential customer over to the kiosk) without using up action video space. In training programs, still frames are the question and menu frames; they are the glue that holds the different segments together and holds the user's hand while presenting important information.

Laying off still frames to a videodisc requires a degree of precision and much planning, but there are tools that make the job easier.

Computer-controlled editing systems allow frame accurate cutting, and electronic still stores allow the grabbing and storing of as many as 800 images at a time. New digital VCRs by Sony and Ampex increase that capacity by a quantum leap, considering that a Sony cassette can hold 90 minutes of live action—or 162,000 frames.

Sound may be added to single frames, using a variety of standardized methods. These vary from black frames holding sound information that "blink" before the image and download the sound into the system to audio hidden in the vertical interval line, which passes through every video frame as the scan beam moves back up to the top of the screen to lay down the next field of video. The sound can range from "telephone" to "FM radio" quality—depending on how much you have to say in a given time.

STILL FRAME SELECTION

When selecting still frames, you have the option of selecting them one frame at a time ("discrete frames") or adding a margin of safety and selecting three or more frames of the same still picture to lay down on videotape. Obviously, still frames will use up less disc real estate than multiples of the same shot. If you have ever edited tape, you already have a good idea why a safety margin is preferred.

Although videotape editing has come a long way and the process is now controlled by a microprocessor, there are mechanical limits to consider.

Remember that a video frame is made up of two video fields. The screen is scanned once, leaving the odd field lines behind, then the scan moves back to the top of the screen using the vertical interval bar and scans down once again, laying down the even lines. Odd and even lines interlace forming a single screen image. One frame is created every 1/30th of a second for a frame rate of 30 frames per second.

The trick in editing still frames is to capture a single frame of video. It is possible to capture one field from a frame together with a field from the next frame. Imagine a frame of a car and a frame of an apple. Mixing the odd field lines of the car with the even field lines of the apple would yield an unsatisfactory result. (See Figure 6.1.) This kind of error is human, not mechanical. Unfortunately, a single, discrete, field-consistent error cannot always be detected until the disc is returned.

There are a couple of exceptions. Both the Ampex VPR-2B and VPR-3 and the Sony BVH 2000 VTRs allow absolute frame and field accuracy. If you put them in the frame display mode and inch them along through the edit with a joystick, they show you both fields when you park on a frame. This field frame mode will save much grief later. Your alternative is to carefully examine every frame by its time code as it is locked onto the screen. Time code is the videotape equivalent of edge numbers on film. Every videotape frame has a time code displayed in the following form:

Hours	Minutes	Seconds	Frames
01:	30:	15:	10

A discrete single frame edit will advance the time code in this way:

```
Frame No. 1 . . . . . . . . . . . . . . . . . . . . . . . .01:30:15:10
Frame No. 2 . . . . . . . . . . . . . . . . . . . . . . .01:30:15:11
Frame No. 3 . . . . . . . . . . . . . . . . . . . . . . .01:30:15:12
```

If there is a large number of still frames to be edited onto a tape, the wear on the editing machine's shuttle mechanism, record/read heads and the editor's mental state will be significant.

Another possibility is to use a frame store device. This device is capable of capturing single, full-field frames and storing them on magnetic disc for later playback in sequence. This ability makes cartoon animation possible and recall of prere-

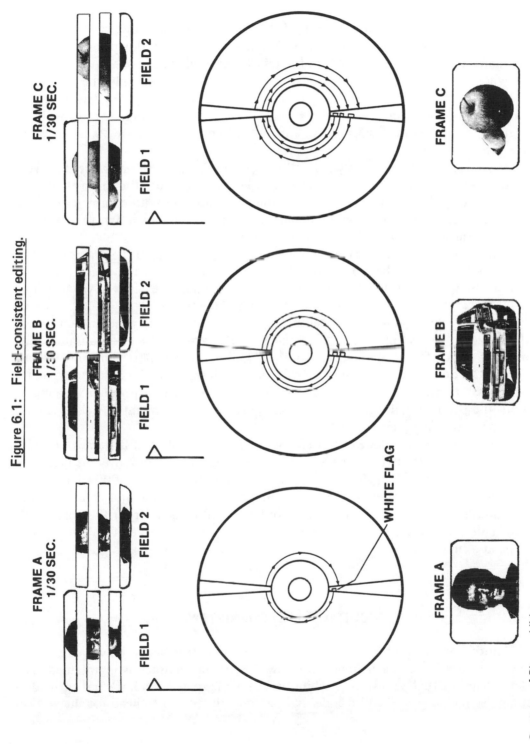

Figure 6.1: Field-consistent editing.

corded stills to be dropped into live programming much like a very expensive slide projector. Frame store devices are available in most moderate-to-top dollar tape editing shops.

To avoid this problem of "wrong field" edits, you can choose not to employ frame accurate editing and go for the safety margin of laying down more than one still frame. Lay down three or more frames of that single image you wish to display and, when you instruct the program to call up this frame, use only the one frame number, e.g., 50007. By doing this, you will have created a cushion by having frames 50005, 50006, 50008 and 50009 also available. You virtually eliminate the possibility of getting caught with a wrong field edit but, of course, you have "wasted" some real estate on the disc.

If you are consistent in always calling up, for example, three frames for every one you require, then the person who is programming your disc from your flow chart will always be able to call up the middle frame number with complete security. This sort of helpful editing on your part eliminates problems down the road. It's called creating a "landing pad."

The industry is so often caught up in ballyhooing the 54,000 still-frame capacity of a 12-inch videodisc that to use that capacity becomes an end in itself. When was the last time you considered showing a 54,000-frame slide show? Any program considering as many as 2000 still frames has a long row to hoe.

Actually, the easiest way to capture many still frames for a videodisc is to use film instead of videotape. Take your slides or flat art, lay them on an animation stand and click off one frame after another. This is the cheapest and most reliable way. On the downside, if you end up with one of those frames badly exposed and it is the *only* frame you have of that subject, you have a real problem. Trying to correct the problem will run the cost back up to where you might as well have used videotape in the first place.

If you are still enamored with laying down a disc full of still frames, consider cost. At an average of two slides per minute (working with consistently well-exposed slides) and producing 120 slides per hour, you could record all 54,000 slides in 450 hours. At an average rate of $350 an hour for editing suite rental complete with editor and assistant, the tab would be about $157,000. Aside from the cost, it can be difficult to convince an editing house that you really want to edit only one frame of tape at a time.

ACCURACY AND CONSISTENCY

Another interesting aspect of still frame use on videodiscs is the way the player knows the location of each frame and how this frame accuracy is maintained. The process requires that you show a white flag. (See Figure 6.2.) No, it's not surrender, just a pulse included in the disc code that instructs the player to head for the vertical interval.

Figure 6.2: White flag.

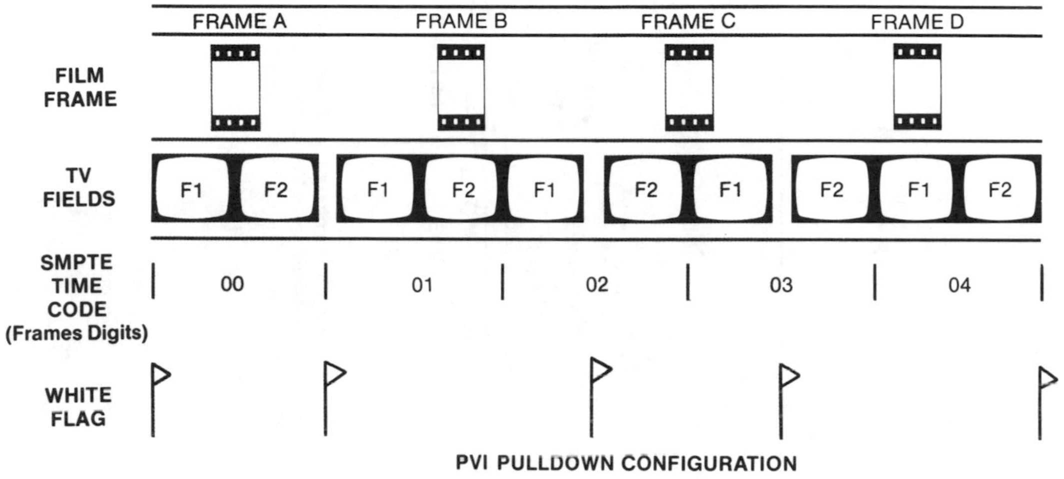

Courtesy of Pioneer Video.

The vertical interval on a CAV disc consists of the wedge-shaped lines radiating from the center to the edge, bisecting the disc's diameter. They represent the black bar that travels from the bottom of the screen to the top carrying the frame address information. They also represent half the diameter of the disc for the tracks that carry the video and audio data. Each track is equally divided into two fields by these vertical intervals.

The white flag pulse is placed in the first field of a video frame, inserted there during the encoding. The player is not smart enough to know which two fields you want to show, so we have to tell it in some way. This white flag, or actually a white video level, is placed in the vertical interval to tell the player to jump back and play those two fields. So when you search for a frame, or stop or step forward, the only way the player knows which fields to pair up is by that white flag.

That flag must be placed so it is "Field Consistent." If you begin your flagging with field No. 1, then you must keep all flags placed at field No. 1. Failing to do this will put the flag on the wrong side (see Figure 6.3) and you end up with a wrong field edit. Actually, you don't even see an interesting hybrid, but as both fields compete for your attention they form a vibrating, wiggling, jiggling mess. This is the importance of field consistency in editing. You can start your flagging at field No. 2 as long as the encoder knows where you started.

To eliminate this problem, allow three frames instead of placing all your bets on one discrete frame.

Figure 6.3: Wrong field edit.

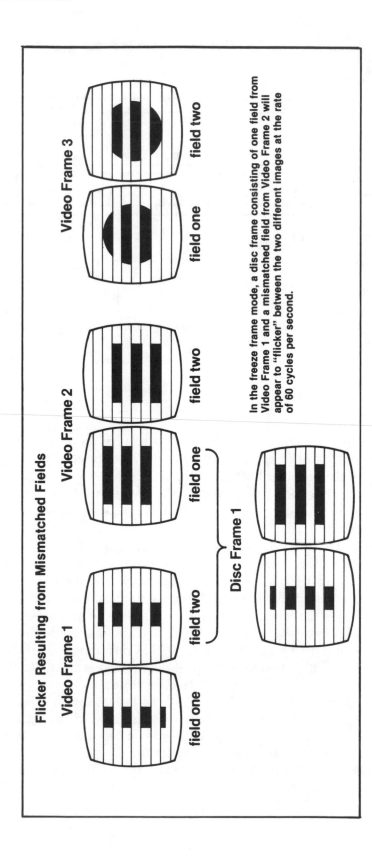

Courtesy of 3M Optical Recording

Flags also figure in motion sequences. Motion picture film is shot at 24 frames per second. In order for it to transfer to videodisc, that 24 frames must be converted to 30 frames per second. The easiest ways to do this are to convert the film to tape, using either a five-bladed shutter on a film chain projector, or to run the film through a Rank Flying Spot Scanner, a machine commonly used for film-to-tape transfers. This transfer is done using a "3-2 pulldown."

A 3-2 pulldown is accomplished by taking film frame A and making it a two-field video frame, then film frame B is held in the projection gate for an additional field (3 fields) so film frame C can come up as a two-field frame. The cycle is repeated: 3 frames—2 frames—3 frames—2 frames, etc. The flags are actually moved to create this 3-2 sequence. (See Figure 6.4.) This cycle repeats once the pattern is established. The result is 43,200 addressable frames rather than the full 54,000 frames. A properly flagged 3-2 pulldown bit of film will show each distinct frame as you step through the sequence. Improperly flagged film will show a jitter. If you find a jitter, you can fix the problem by repositioning the flags to maintain the 3-2 pulldown.

In video, if frame jitter is discovered, there's no easy fix. In film you are actually transferring individual, unique frames of film whereas in video, you are *scanning* an image to create a frame.

You will usually only be aware of this phenomenon if you choose to step through still frames of a motion sequence to create "slow motion," which is not the best way to create this effect. If you want slow motion, shoot it that way in the camera. There are slow motion video camera systems as well as the wild motor on a film camera, or specialty high-speed cameras. Today, many film cameras offer 30 frames per second with sound sync to make the transition from film to tape easier. Most made-for-TV films are now edited as videotape.

STILL FRAME DESIGN

The frames themselves require a bit of planning as far as content and formatting are concerned. Still frame sources include camera-ready text or text from a video character generator. Slides can be used as long as they are cropped to the video aspect of three units high by four units wide. Still images can come from film clips or flat art. If your still image suffers from the blahs, most post-production houses can posterize, crop, enlarge or even turn it into a mosaic. One caution when using electronically generated text from a character generator: reduce the color saturation to about 60% or 12.5% on the vector-scope; otherwise, the color will be too intense.

When designing the still frames, there are definite considerations.

A good rule for text frames shown to groups is to allow no more than 12 lines of text per frame, with 25 characters per line. Always consider the largest viewing audience for your frame. If you are sure only one person will be watching, the rule can be expanded to 15 lines of text with 50 characters per line. If you plan on unusual or particularly ornate text, reduce both these numbers.

Figure 6.4:   A 3–2 pulldown.

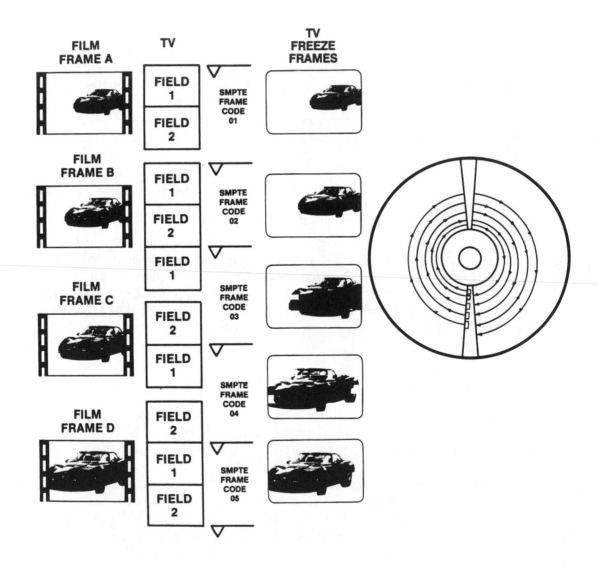

Courtesy of Pioneer Video.

Images combining text with background pictures can cause confusion for your audience. They must always know where to look and be able to diferentiate between foreground and background. Text can be enclosed in solid color boxes, or a 20% gray screen can be overlayed on the background image.

Another temptation is to provide too many menu choices for the user on a single screen. Six is a good maximum; simply use more screens for more choices. If some "global options" are offered (such as those that follow), they should be offered judiciously throughout the program and not repeated on every screen.

Index—Return to the main index (menu).
Repeat—Replay a portion of a motion sequence.
Skip—Skip the rest of a motion sequence.
Exit—Take the user to a still frame that explains how to exit the disc.
Help—Offers supplementary information to the user.

SOUND

One unfortunate side effect of using stills in videodisc programs was, until quite recently, their silence. There was no way to provide sound for them because the disc was spinning under the laser, which was frozen to a single track.

This problem was solved by encoding "compressed audio" in an area of black just prior to the still frame. During this process the screen "blinks," then the audio is played under the slide. Another version crops a section of black from the actual image. The top or bottom 20% of the screen goes black as the audio message is played.

A caution is offered here. Sound with still frames is still in its infancy and the editing tends to be a bit of a nightmare. Until we can automate the editing system that merges audio, video and digital data, sound with stills has a long way to go.

A company called EECO in Santa Ana, CA, produces an analog sound compression technique that is light years ahead of the current analog-to-digital-back-to-analog systems used by Sony and Pioneer as far as *ease of use* is concerned. The EECO's sound quality suffers a bit, but only in comparison. At present, the EECOder is a proprietary system licensed through 3M in the United States. Pioneer and Sony do their encoding in Japan, so you have to allow for the delay.

The Pioneer sound with stills and data system (SWSD) offers three audio quality choices depending on how much time per frame is desired. The first offers 3.6 seconds of "telephone quality" per frame (35.2 hours per side). The second steps up a peg to 1.8 seconds of "AM radio quality" (17.6 hours per side), which is really quite acceptable. The third choice tops out with 1.2 seconds of "better than AM radio quality" (11.7 hours per side). This last audio choice plays back excellent music and voice.

A frame is coded to let the Pioneer SWSD (Still with Sound and Data) encoder know that some of the following frames are encoded with compressed digitized audio data. The SWSD encoder reads that digitized audio data and stores it in its digital memory. The player then displays the frame while the SWSD processor decodes and cleans the data from its memory. The screen goes blank for an instant just before a still frame is displayed while the processor reads the data from the preceding frame. This instant of blankness is the reason Pioneer has dubbed this the "Blinking Method."

A single disc side can store up to 20 hours of audio, describing some 4500 still frames. About 100 typical sound film strips can be accommodated on each side of a videodisc. Digital storage capacity of a videodisc can be quite awesome, allowing the data from 1000 floppy disks to be stored on one videodisc side. (See Chapter 9 for further discussion of alternative technologies.)

MARKET APPLICATIONS

The market for still frame applications is almost totally confined to catalogs and archive storage. One of the more interesting and successful applications is the Patsearch system marketed by Pergamon. Videodiscs are used to store patent drawings and data. Patsearch is offered to lawyers who deal in patent search law. The system requires lawyers to have a computer, such as an IBM PC, and a videodisc player. They receive videodiscs and software that is used to locate the required frames. The system is expensive, but is far cheaper than flying to Washington, DC, and pouring through the patent files for days as is the normal patent search procedure. The discs are updated periodically.

Another successful use of videodisc still storage is broadcast television. Because the videodisc puts out direct broadcastable color, television stations use the disc to store station logos, beauty shots for weather, news and station breaks. These can be mixed with motion teasers and opening graphic animation sequences.

CONCLUSION

Too often, interactive video designers become enamored with the motion sequences in their programs to the detriment of the still-frame capabilities that are unique to the videodisc format. This chapter has looked at some of those applications.

As we move on to discuss program planning, design and production, you are encouraged to remember that still frames are as important a consideration as live action in your interactive program. They should be accorded the same care in your planning.

7 Interactive Videodisc Design and Production

After plowing through all the hardware, software, computer jargon, video jargon, authoring languages, a helping of history and a look at applications, you are ready to address the task of designing and producing an interactive video program.

First, a word of caution. How many people bought home computers when they first appeared in the discount stores, then hunted around for something to compute? Approaching videodisc technology can offer similar pitfalls.

Too often, the *idea* of interactive video overcomes the *reality* of interactive video. It is so easy to imagine that the videodisc and its touted features can be the panacea for all your training problems or point-of-sales blues. The hardware is flashy; state of the art technology and a good demonstration disc can start the salivary glands pumping. Take a few deep breaths before plunging in.

An effective, interactive videodisc requires planning and organization akin to a complex military campaign. It requires time and a thorough understanding of the objectives to be achieved. It demands an analysis of the tasks under consideration and will only be effective if every step is documented completely. (See Figure 7.1.)

The guidelines for interactive videodisc design have been compiled the hard way, by individuals who have been over the jumps, made mistakes and passed on what they have learned. Every author on the subject puts his or her own particular spin on the ball, but there are many islands of common ground. This chapter will look at those islands.

GOALS AND OBJECTIVES FOR EDUCATIONAL PROGRAMS

One doesn't lurch forward into interactivity. One proceeds with a plan in mind. You must have a goal, a clear vision of the reason for creating this expensive, exhaustive production. The only reason for an interactive video program is to change someone's behavior—to teach a lesson, to enlighten, to cause a response and to have the user benefit from that response. Benefit how? Teach what? This is what you must decide at the outset. To help with this first step, successful pioneers have given us a sort of toolbox.

Figure 7.1:   A sample production sequence.

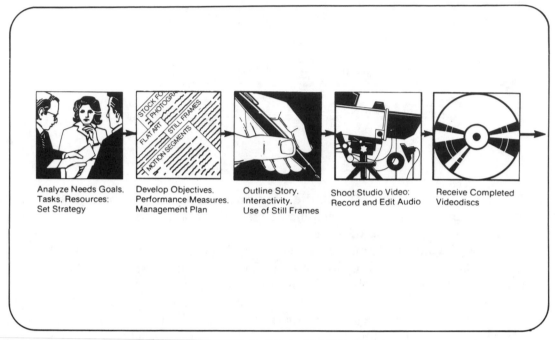

Analyze Needs Goals.
Tasks. Resources:
Set Strategy

Develop Objectives.
Performance Measures.
Management Plan

Outline Story.
Interactivity.
Use of Still Frames

Shoot Studio Video:
Record and Edit Audio

Receive Completed
Videodiscs

Courtesy of WICAT Systems, Inc.

Analyses have been developed by different designers and called by different names. The book *A Practical Guide to Interactive Video Design* by Nicholas V. Iuppa (Knowledge Industry Publications, Inc.) is a particularly excellent volume on the subject. By combining Iuppa's observations with guidelines from other experts we can assemble an easy-to-follow road map.

Goals Analysis

Goals analysis is the first overall examination of a proposed program. You must evaluate the project's scope, how much time you have to complete it, how much money has been allotted to get the job done and if the goals are consistent with the budget.

This beginning must be bound up in paper—a clear document defining each step. All parties to the project must have their say. Possible alternatives and modifications of the original plan must be considered, and the media to be employed must be determined, whether it be videotape, film or slides. The sound track must also be considered. Will it be bilingual and use audio track two for a second language? Will the chosen media be compatible with the message and with the budget? Raise all these questions and considerations to bring any misconceptions or preconceptions to the surface. Only when the program's goals have been carved out in this way—a broad-brush plan—can you proceed to the next step, or steps.

Front-End Analysis

This description originated with Joe Harless in *An Ounce of Analysis Is Worth a Pound of Objectives* (Harless Performance Guild, Newnan, GA). The approach asks if the proposed interactive program will achieve the desired solution.

For instance, if you want to train people to use flimet grinders (a fictitious device) more effectively, you must determine that people are currently using them improperly. One way to investigate is to approach flimet grinder operators on the line and observe them grinding flimets. If they can do the job successfully while you are observing, then perhaps the problem is not that they cannot do the job, but that they do not *want* to do the job.

In this case, front-end analysis shows that more training isn't needed, but better motivation or a better work environment is necessary. If you overlooked this step, you could stick out your neck, plunge into an interactive program that is teaching people what they already know and miss the real cause of production problems.

What do you do if you discover that training is necessary? What if, as the one chosen to produce this epic, you don't know a flimet grinder from a herring stretcher?

What if you are trying to peddle flimet grinders and your front-end analysis tells you that point-of-sale interactive kiosks will put potential customers directly in contact with your top-of-the-line grinder? How do you attract and hold the interest of your buying public?

Task Analysis

This is where you do your homework. What exactly are you going to teach? What response are you trying to elicit?

You have to define the skills you are going to teach if your goal is a training program. To do this, a careful study must be made to document these necessary skills. If you know nothing of flimet grinders, then every scrap of information relative to the flimet world must be absorbed. Particularly important is that same step used in our front-end analysis: carefully observe an expert flimet grinder operator and note *every* step, especially tricks used by the operator that have been learned by experience. Often, the manual that came with the machine has been "improved upon" by the expert grinder. Draw up a complete list. If you need more answers, talk to the expert.

If your project is sales-oriented, product facts and descriptions must be learned. Who uses a flimet grinder? Is it competitively priced? What are its features, its benefits? How can the potential customer be made aware of the mighty flimet grinder? Document all these facts and figures.

You are creating a list of tasks to be taught based on careful analysis, or a definition of the message your interactive program will convey based on an equally exacting examination.

This is probably the most important part of developing an interactive program. Be incredibly fussy, because the next step looks back at your original objectives, but this time in terms of "action."

We've examined our criteria for getting into this project, then we dissected those goals in the form of analysis of the required result. Now, having defined our path to the solution, we have to look at a more specific goal: "The student (potential customer) will be able to _____ ."

We must define how the user's behavior will be changed: "the user will be able to understand the theory of relativity" or "the user will be able to rent a lawn-mower." The more solid this statement can be in terms of realization, the better. For example, "the user will understand how to rent things" is wishy-washy, as is "the user will develop a grasp of Einstein's theory." The action terms "understand," "become familiar with," or "appreciate" are not strong enough indicators of the purpose of your project for the product group, education board or marketing division that will sit down to examine your premises before handing you vast sums of money. Generalizations at this stage do not show your grasp of either marketing or educational theory—the tools you will use to attain your goal. Carve the behavioral objectives in stone.

If this seems a tedious path to take if all you want to do is to teach some children basic addition, you are right. We are laying out steps to cover the complexities. Take what you need but remember: even simple courseware is the result of much study and knowledge of your audience.

Teaching Method

In order to teach your audience the desired skill or skills, the following basic teaching concepts can help you plan the lesson.*

Prerequisites

The "why" of the lesson should be stated up front, establishing the context and explaining the language, basic concepts and expected result.

Step Size

How much material should be fed to the student at one time? The best method is to go easy—two or three ideas at a time before actual practice exercises begin.

Sequence

Sequence pertains to the order of presentation. The sequence does not have to be linear or chronological, but should indicate what strategies will be used to maneuver the student through the program.

---

*Note: The basic teaching concepts discussed in this section are taken, in part, from "The Performance Problem-Solving Workshop" created by Joe Harless (Harless Performance Guild, Newman, GA).

Simulation

This is a broad-brush approach to demonstrating the degree of realism involved in presenting the material to the student. Getting bogged down in details here is not necessary or desirable. Details can be added later as the design is fleshed out.

Feedback

Here we get to the kernel of interactivity. How will the student be given information after responding to your program? At this point, a close tracking with the student's possible responses is recommended.

Exercise

A sample exercise would be helpful to demonstrate how the information will be presented, how it will be prompted and, finally, tested. A close simulation of the skill should be a part of this exercise.

Practice

This strategy falls into two possible categories. The first is isolated practice in the form of drills, or working with a simulation without any distractions from the program in the form of prompts, questions or branching feedback. The second is integrated practice, placing the practice in the context of the real world where situations change unexpectedly. Here new problems can develop from the solution to a *part* of the original problem. This is where distractions (in the form of branching, program intrusions, and so on) create a feeling of performing the skill in a real-life situation.

Design Document

After all these considerations are set down and understood, it is time to produce a design document. This should be prepared in two forms: one is for the client, showing the broad design in the client's terms; the second is for producers, designers, writers—the folks who will have the responsibility of producing your grand design. These documents should show depth so everyone connected with the project is "reading from the same music." Depth, however, does not mean a doctoral thesis on interactive video. Keep your documentation concise.

Documentation can be arranged so that it fits your own needs, as long as the worksheet can be photocopied and paged through as you would read a book. The elements themselves can be broken down into four or five basic columns. The work sheet lays out the following from left to right:

Event No.   Kind of Event   Objective   Content   Visualization

The first column represents the number of the event. The event can be broken into a number and a sub-number, or letter, to group subsets of an event under that event (i.e., 1.A, 1.A.1, 1.A.2). Next comes a description of the kind of event we are planning. Is it a question, some student feedback, a menu or a description of what happens when the student gives a correct or incorrect answer? The reason for, or objective of, the event is next in line. Some examples would be: "The student will be able to recognize the magneto fault," or "The student will be able to trace the magneto fault to its source."

The content of the event is then summarized with a few lines: "The magneto grinds to a halt with a grating sound." Finally, we list a description of the visualization of the event—what the student will see on the screen: "Motion sequence followed by a freeze-frame."

GOALS AND OBJECTIVES FOR POINT-OF-SALE SYSTEMS

Taking the principles we have discussed and relating them directly to a point-of-sale (POS) program requires little reworking. In the case of a commercial interactive video plan, you are presenting a product or a service in order to involve a potential customer in your design. If the disc design is successful, the potential customer will make decisions based on the effectiveness of your interactivity.

It is critical that, because of the uncontrolled environment in which your POS is presented, you target both your audience and your location. In a classroom situation, students are paraded into the learning experience by instructors. The situation is structured, and the audience is, usually, captive. With POS programs, however, if the users become bored or confused at the outset, or perceive that they are in for hard work before they get to the meat of the program, they walk away. Worse yet, they complain to the store manager or ship off a nasty letter to your client. The POS display is cast into the public arena and there are no safety nets.

Determining the Audience

You must ask definite questions when trying to determine your user audience:

1. Who is the actual audience? Women who shop in a mall? Travelers passing through an airport? Shoppers in a store with definite purchases in mind?
2. Is there a secondary audience? Will the program also involve the curious, or potential customers outside the original audience target?
3. What are the expectations of the users as they approach the program? Are they familiar with videodiscs? Is a level of understanding of the program a

prerequisite, or can a person unfamiliar with POS displays feel comfortable with the design?

4. Should any additional consideration be given to a mix of sexes or demographics among the target audience?
5. Is it necessary to use the second sound track to provide a bilingual program?
6. What, if any, should be the mix of entertainment or visual stimulation with the hard message?
7. Will a user be returning to this POS display? Should there be a way for the experienced user to shortcut into the actual services without wading through the program introduction?
8. Are comparisons being used that will give your client's competitors free publicity?
9. Have you let technological "bells and whistles" get in the way of the message?
10. Does the program make learning about your client's product or service an enjoyable experience? People retain more of a message if they become actively involved with a program than if they are simply presented with images and ideas in a passive context. Resistance to an interactive program is less if the experience seems to be useful.

PROGRAM DESIGN

In a recent traveling seminar on interactive video conducted by Pioneer, we compiled a list of "Questions a User Should Never Have to Ask." The first batch had to do with clarity of program content:

- What should my answer have been?
- Why was my answer wrong?
- What does that have to do with what you told me before?
- What do you mean?
- For example?

Another set of questions represented a cross section of program design errors:

- How did I get here?
- What am I supposed to do now?
- Where am I?
- What other choices do I have?
- What happens if I press. . . ?
- How do I get back?
- How do I get out of this sequence?
- Can I review that?
- Where am I supposed to look?
- What should I do if I do not understand?

It is amazing what an audience will endure to obtain information it really wants, but why make it difficult? Lynn Yeazel, a long-time pioneer in the development of videodisc systems, shared an experience with the group at the seminar that demonstrated how much people will endure for needed information.

A gentleman approached Lynn to produce a program that involved some complicated scientific ideas. He brought along his charts and graphs and a corner of a studio was set up according to his wishes. He stepped in front of a single camera. Hours later, the program was finished. It was of embarrassingly poor quality, but the client was more than satisfied with his stand-up monologue. Some time later, Yeazel ran into another scientist who had seen the program. When asked if he had made the tape, Lynn apologetically admitted that he had. To his surprise the scientist gushed, "That was a great production. That guy had information we needed in the field. Everybody's trying to get a copy!"

The information was important to a particular group—more important than the inept delivery. Content is critical in any production. While children might appreciate a somewhat round-about, "candified" approach, mature adults have certain learning prerequisites. They appreciate immediate feedback to a response and some sort of remediation such as "No, you forgot," or "That's correct. Have you considered . . . ." The feedback should contain some variety appropriate to the user's response and not just a constant repeat of previously presented material.

The user should be able to skip material that has become familiar and proceed with the next level of the program. Material should be equally accessible for review and reference.

In particular, a program should be easy to use. Whatever you design, the old cliche "user-friendly" should be painted in large letters and nailed to the wall of your office.

An interactive video program must *not* pass judgment on the user. Responses such as "I think you really know better" or "Good grief! Five wrong answers in a row! Press the 'R' key for remedial" are clearly not acceptable. Also, excess enthusiasm is inappropriate—especially when it is not warranted: "You got that one right. Let's try another!"

Even small design elements can become irritating. If you have ever worked on a program that kept requiring you to press *two* keys to initiate a response, you understand.

This is lazy programming. Even elementary BASIC language offers a "Get" statement that goes into action as soon as you press "Y" or "N" to branch to another part of the program. An input statement requires the "return" function to respond. The use of the input statement can get pretty old for the user after a dozen or so two-key responses.

But in your effort to maintain user-friendliness, don't overlook the unexpected or surprise twists that add interest to your design. Small touches can give an interactive program some class.

The Flowchart

Designing flowcharts is a discipline that is essential to producing an effective, organized interactive video program. To anyone who has had a brush with computer programming, a flowchart is a familiar sight. (See Figure 7.2.) Digital programmers' charts present not only a guide to the program's logic, but a myriad of technical details that are meaningful only to other digital programmers. This degree of detail is not necessary to show the expected flow of your interactive program. It *is* necessary, however, that you go through the exercise of creating a chart for two reasons.

First, a flowchart helps you solidify your own plan for the program. The visuals, responses, feedback, motion sequences, still frames and branching can all be visualized. If there are any flaws in your branching, they will materialize during the exercise. A plan that exists in written form will often take on new life in flowchart form as new possibilities take shape.

Second, the chart will serve as a guide for A/V producers, designers and encoders who will work on your design. The standardization of many flowchart symbols adds clarity.

In a linear program, there is no need for flowcharts. The sequence of events leads from A to B without branching. In interactive video, the flowchart reveals all the paths open to the user as well as the visuals and sounds encountered, remedial activities, default values and even user tests.

Each element of an interactive program is represented by a flowchart symbol. The symbols are arranged according to a standardized format, each linking into an easily followed map. The maps themselves are usually separated into three categories:

1. A basic map for the writers and clients that shows the program's flow with a broad brush.
2. A tighter version designed for the production people who will have to implement the program's structure with hard decisions.
3. The nitty-gritty flowchart designed for programmers who end up with the finished product and must convert the elements into computer code.

Of course, there are many variations of the basic flowchart idea. Some designers use a code to label sequences consisting of numbers and letters within a symbol. Others adopt slightly different interpretations of standard symbol shapes. You can pick and choose your symbology, but you must use flowcharts. It is a subject best shown by example. A catalog of interactive flowchart symbols is illustrated in Figure 7.3.

As you can see from the catalog of symbols, the flowchart is a symbolic road map that provides an overview of the program's structure. The sample chart furnished courtesy of Pioneer Video (see Figure 7.2) is for a Level 2 program. It lacks the symbols that are associated with Level 3 productions such as text overlay and manual operation (see Figure 7.4).

Figure 7.2:   A sample annotated flowchart.

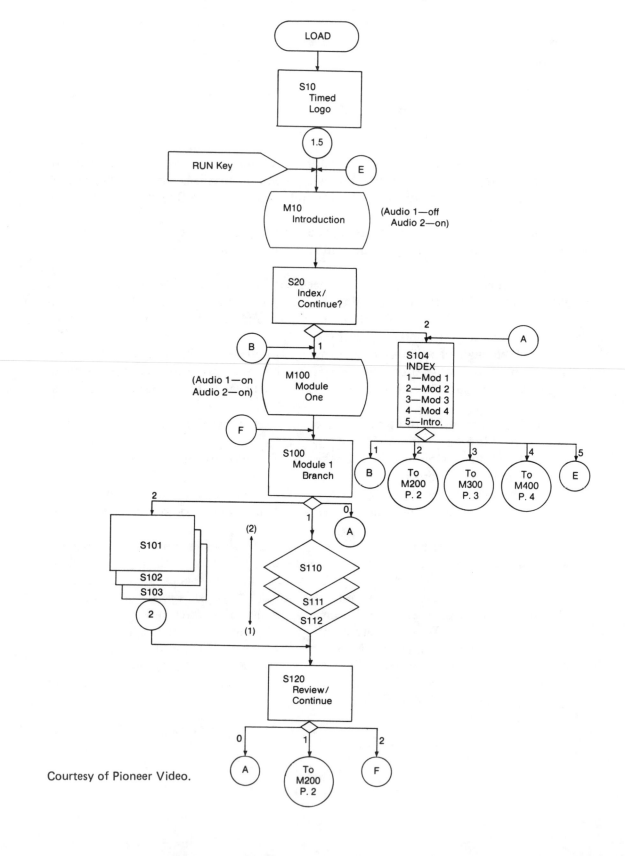

Courtesy of Pioneer Video.

Figure 7.3:   Interactive flowchart symbols.

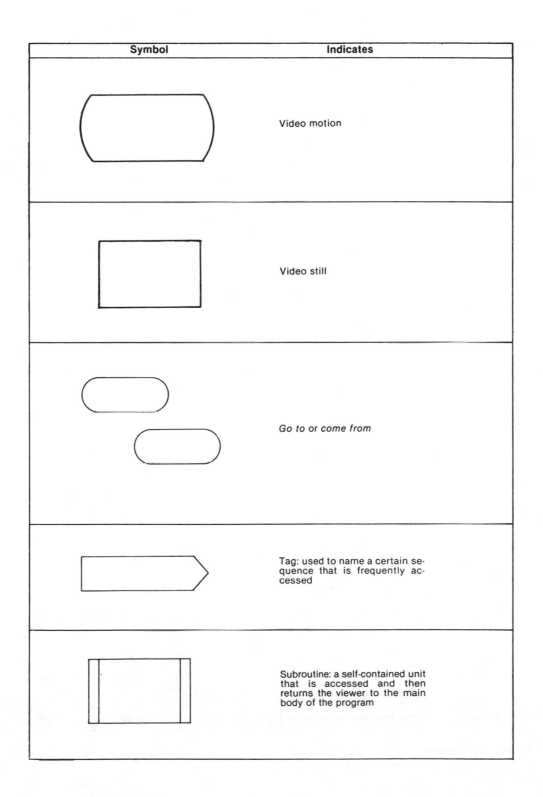

Symbol	Indicates
	Video motion
	Video still
	Go to or *come from*
	Tag: used to name a certain sequence that is frequently accessed
	Subroutine: a self-contained unit that is accessed and then returns the viewer to the main body of the program

Figure 7.3:   Interactive flowchart symbols (cont.).

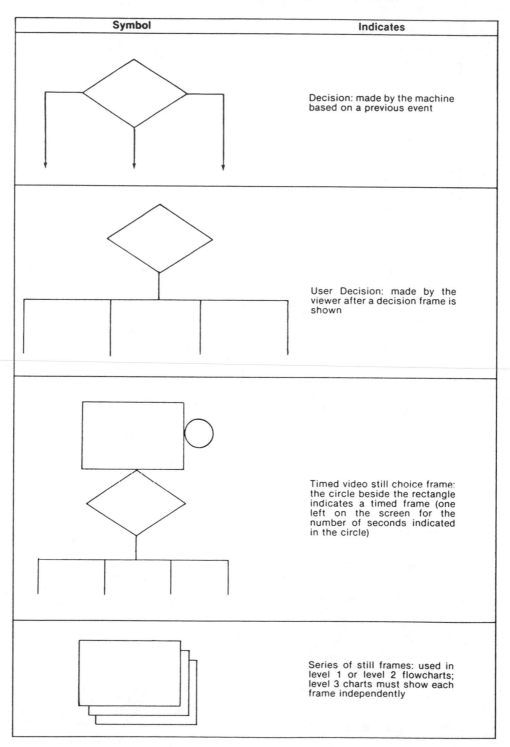

Source: Reprinted from *A Practical Guide to Interactive Video Design,* by Nicholas V. Iuppa (Knowledge Industry Publications, Inc.).

Figure 7.4:   Level 3 flowchart symbols.

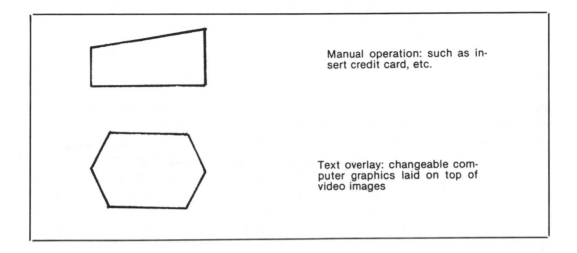

Manual operation: such as insert credit card, etc.

Text overlay: changeable computer graphics laid on top of
video images

Source: Reprinted from *A Practical Guide to Interactive Video,* by Nicholas V. Iuppa (Knowledge
Industry Publications, Inc.).

The single most important aspect of the flowchart is *clarity*. While the catalog
of symbols is fairly comprehensive, the only standard is that there is no "standard"
set of shapes. New symbols and coding evolve as the need arises. To burden a programmer or designer with a chronicle of shape architecture to the detriment of
understanding is to defeat the purpose of the flowchart.

PROGRAM PRODUCTION

After selling the idea, planning, documenting and creating flowcharts, your
opus is ready to spring from the paper into reality. Pause a moment. This is the
point at which the financial clock begins to wind in earnest.

In-House Facilities

If you are operating with an in-house production service, you have a duty to
uphold your standards at any cost, especially if the facility has never produced an
interactive videodisc program. The way is long and arduous, requiring scrupulous
production values and facilities capable of professional results. Always keep in mind
that your program will be judged by users—whether students in a training program
or the public at large in a POS situation—by the standards they have become used to
in broadcast television.

I worked for Motorola Communications for eight years. I began as a writer and
still photographer and worked up through film to video production. The greatest

hurdle I encountered when I worked at an in-house facility was the corporate "video" mentality. The in-house production facility is often considered to be the "cheap spread," while outside services are the "butter." Simple training and sales tapes are the grist for in-house production. These are easy to grind out since the costs are based on "funny money," paper budgets showing little or no markup and utilizing salaried employees instead of day-rate professionals or high-ticket editing shops.

Fortunately, the cost of producing professional video has come down and companies are finding the ability to produce their own tapes is easier than anticipated. But even the most sophisticated in-house systems often find they are relegated to "basic programs." Industrial video is generally created for a captive audience in much the same sense as many educational programs are produced for students.

Salespersons are herded into "training sessions" and seated in front of a video monitor. Budgets often preclude production standards that are common in television, but these linear productions are not supposed to be slick, just functional. How many times, as a child, were you led to the assembly hall, or suddenly found the classroom shades being drawn, only to confront something called "The Romance of Wheat"?

Salespersons and other trainees suffer the same fate, peering at the in-house creation squeezed from small, artificial budgets under artificial deadlines. The product is often artificial video.

Interactive video cannot take such shabby treatment. If all the planning has been completed on paper according to the best standards, then that planning must be pressed home according to the same standards when the cameras begin to roll. It's very important that the decision makers who have approved the script and all the other benchmarks along the way understand this commitment.

Interactive video demands new ways of seeing and approaches quite different from linear productions. For example, videodisc resolution is critical. The images are sharp and the sound quality can be in stereo; the resolution is far superior to videotape. When a VHS duplicate tape is dubbed from a 1-inch master tape, some resolution is lost and with that loss, some flaws can also be diminished (glitches, drop-outs, and slight color shifts). The videotape process is quite forgiving. Discs, however, hold up a cruel mirror to slack production values. What you produce is what you see, warts and all.

To create interactivity, branching leads to options for the user. These options may be variations of a process—the right and wrong way to do something. A scene may have to be shot many different ways to accommodate these options. Each variation has to stand on its own and not rely on any support from the other options. The user's relationship to the program must be seamless. A scene must appear to be a natural response to a user's choice. Any flaws in production technique disturb the rapport established between the user and the program.

Video must also coexist with computer programming. There are so many facets to be taken into account that many persons must accept part of the divided responsibility of production. You cannot simply throw the ball to the company video production people and jump back. An overall production manager must be named to hover above the mechanics, the creativity, the digital programming, the production sound stages and locations—all the vital activities required by the interactive medium.

When it is pointed out that this kind of administration will also help maintain budget scrutiny, the manager's job is not very hard to sell to the corporate decision makers.

Another production consideration is the fact that many in-house and quite a few outside post-production facilities are geared to motion sequences, but are not equipped to handle a considerable quantity of complex still frames. Recording discrete video frames requires precise editing machinery, a method for still-frame storage and a very good camera. If the still frame is derived from a projected transparency or flat art, the camera's ability to resolve detail and a broad range of contrast is critical.

Outside Production Houses

If your in-house system is not capable of producing this level of broadcast television standards, or you must rely on outside facilities for production and post production, what can you expect to pay? How can you control costs?

There are some guideposts to help stretch those dollars and to avoid ugly surprises once the project is under way. Again your salvation lies in research and understanding both real and perceived needs.

Research begins, or rather continues, with a canvas of the local production houses. Every major city has some sort of videotape production facility. If you are living in the "hinterlands," then you have two other possibilities: the local television station or a college or university.

TV stations, like newspapers that handle job printing for wedding invitations, often provide video services for customers who advertise on that station. They have rates for their facilities, editing and crews. In communities served by cable, cable access studios are frequently included with the cable package.

In the same way, local colleges have often built a sophisticated video facility with the help of endowments and grants. The chance to work on an interactive project is often a good lead-in to engage the interest of members of the academic community. They may have already produced a program of their own and, in that case, their help would be of great value.

With these alternatives, be especially scrupulous in your requirements. Low-budget operations often rely on trainees or interns. While these folk are full of enthusiasm, their skill levels are not necessarily at peak. By using care and clear documentation, however, even this low end of the production scale can be utilized.

If your headquarters is among the civilized, shop the Yellow Pages and make some phone calls, weed out the obvious and narrow your hunt to the most promising. Once you've settled on your production house, paper work again comes into play. You must develop a contract that spells out exactly what is required, this includes the objectives of the program and the final shape of the master tape elements. Be very specific and detail an incremental payment procedure based on approvals of work accomplished. Most ethical houses already use this check and balance method. It protects you and also gives both you and the production house a solid understanding of what is required to get the job done.

Costs

One of the most important considerations at this juncture involves production costs. A few suggestions are offered, gathered from a few years experience in all phases of production.

A videodisc master tape takes about the same time and expense to make as a standard linear tape, except for a few additions discussed below.

Development takes more time because of branching and flowchart requirements. Always remember that an interactive program is tied to computer technology, not just to edited sound and visuals.

Edit Decision List

The compilation of the edit decision list for an interactive tape requires more time than for a linear program. An edit decision list is just that—a list of the edits, in sequence, compiled for the editor. Each edit is marked with an "in" point and an "out" point. The list can be compiled in two ways.

As with the frame numbers on a videodisc, each frame of videotape is identified by "time code." This code is generated by the camera or can be added to the tape later. The frame locations are marked by a number sequence showing hours, minutes, seconds and frames. The code looks like this:

15:20:30:06

When the camera footage has been finished, the tapes are dubbed over to either 3/4-inch or 1/2-inch VHS (or Beta) and the time code is "burned into" a window on the tape so when the images show on the screen, so does the time code. The producer can select time code in and out points for each edit by using an offline editing system that transfers the selected edit to another tape. As the edits are strung together on this tape, a "rough cut" of the program is produced.

On sophisticated systems, each edit is transferred to a punch tape or floppy disk. When the scenes have all been edited, a record of the edits is stored. By playing back the tape or disk into an "online" editing system, the offline decisions can be automatically recreated with the original camera footage.

For a 10- to 20-minute videotape, this offline editing process can take an average of about 16 hours. More time may be allowed for the offline edit because that is where many of the creative editing decisions are made. With an interactive program, this time will at least double due to the branching and constant reference back to the flowchart.

Another, and cheaper, method of compiling an edit decision list is to create a "way" offline edit. This list can be created with a standard four-head VCR that can freeze-frame and that has a remote control with pause and step frame. The window-dub VHS cassettes are rolled through the VCR and the same edit decisions are made, using the time code. This time, however, the in and out points are simply noted on paper and this paper, along with the master reels, is given to the editor at the online session.

Digital Programming

Another function requiring additional time is digital programming. This is unique to interactive videodiscs and tape. Computer programming can proceed, following the flowchart, at the same time as video production. You must allow for a few days extra time, however, for the inevitable debugging and verification that will ensure that the videotape and computer program are truly in sync.

This additional time spent matching the computer program with the video segments also costs additional dollars, generally in the range of $100 an hour for offline editing and $300 to $450 an hour for the online work. You can see why it is important to get it right in offline.

Sound Stage vs. Location Shooting

The clock is really ticking once the cameras are turned on for the first time. When developing the script, consider how many sets, or locations will *actually* be needed. The penny pinchers see the sound stage as a huge money hole, with cameras, lights, sets, carpenters, grid monkeys and rental time amounting to a hefty chunk of the budget. Locations, on the other hand, are *real*. The walls, trees and other background objects are already there, and the sun provides light. Just set up a camera and let the good times roll. Wrong.

In truth, the first budget pass gives the edge to location shooting. However, once the crew has arrived on the location and shooting begins, there are other elements to consider. The first is sound.

Sound can create problems when you try to match shots. For example, if an actor is speaking and a car is heard in the background, then a car has to be heard in the background if a segment of that shoot is redone. You have to shoot a matching shot that cuts from that bit of dialogue or the shots won't match. This is why sound technicians always wait until there is as little outside noise as possible. These sound delays can hold up shooting considerably.

Another problem is light. In a studio, *you* control the lights, their direction and intensity. Mother Nature has other ideas. The sun moves, clouds show up at awkward moments and shadows that were no problem initially become bothersome as time passes.

The initial savings in hardware and sound stage rental that is gained from shooting on-location are gobbled up by the "Time Bandit." The major question to ask is how important is the actual environment for a scene's veracity. Location shooting can add immeasurably to a production's quality if major sets cannot be afforded, but cost alone should not be the deciding factor. Creative need is the most important criterion.

Sometimes, locations can be suggested by the judicial use of stock footage or cuts from previously produced tapes. Stock tape footage is harder to come by than stock film footage. In either case, the material is very expensive and is doled out by the second or by the minute. In any case, review all footage to be used *before* you get into online editing.

Talent and Crew

Once you have begun taping, you should have a final script in hand and, preferably, an organized storyboard showing key scenes, at the least. Changes and rewrites with tape rolling are evidence of shabby planning and preparation. Once the talent has rehearsed and learned the lines and blocking, retakes and rewrites will cause time problems that can add up to great expense.

Talent is another cost consideration. The first knee-jerk reaction to hiring professional talent is to go for a "name." This is a tremendous luxury, if your budget allows for such an expenditure.

If you are not careful, "name" talent can also become a distraction: people on the set may want to talk about the talent's last film; a "star" personality can get in the way of your message; and some "names" may be used to VIP treatment.

When paid talent is involved, be sure all tape is reviewed as soon as possible after each day's shooting. With videotape, it is possible and advisable to review takes as soon as time permits. Talent is expensive, but having to call back talent to set up again and reshoot is much *more* expensive.

This also holds true for crews. A crew is often assembled from freelance professionals. They allot a given amount of time for the schedule. Once they have been released, they are free agents once again and probably have consecutive bookings lined up. If you have to call back a video technician or lighting cameraperson because of "changes" not planned for in the beginning, you may lose several days in your shooting schedule. Such days are very expensive.

To minimize surprises, it is best to have the client, or primary decision maker, on the set during shooting to approve all takes on the spot. If this is not possible, the client should provide a subject-matter technical expert to make sure the nuts and bolts go together correctly.

Animation

Another money sponge is animation, in any form. Cell animation, as in the Disney classics, is horrendously expensive because each element in the cartoon must be hand-drawn and combined in an animation camera. The result is beautiful, but there are easier and infinitely more cost-effective ways of achieving the same effect.

Computer-generated graphics have matured to the point that they are at least as good as cell animation and are inexpensive enough to fit most pocketbooks. Costs can range from as little as $1500 for a simple animated logo to $1000 a second for a complex sequence. Be careful when budgeting for computer-generated graphics. The cost of different systems varies widely.

First, the sequence must be designed on paper, and the elements created as combinations of moving shapes, or polygons. The elements must then be choreographed to create the animation. These steps cost time and money. The final phase—choreographing the animation—can add surprise costs if the movement must be laid to the master tape one frame at a time. Some animation systems permit frame compiling in real time. Inquire about this when you shop for an animation house.

Another alternative to either of the two systems is the use of slides. I have used slides, projected in pin-registered mounts, to great effect. At Motivation Media, where I produce video programs, we have eight designers who have amassed considerable experience in producing multi-projector slide programs. This expertise has been put to use in our Video Systems Division to create some very sophisticated animation—at a savings over computer-generated production and at a far cheaper cost than cell animation. Make certain that all animation elements are approved *before* production.

Organizational Tips

While in the planning stage, there are a number of organizational tips that help produce a smooth and professional-appearing program. By considering design elements up front, time can be saved during editing when bad organization becomes evident and fixes must be made—fixes cost money.

Consider the finished, approved and annotated flowchart to be the editing bible. Together with the edit decision list, the flowchart is a clear path through your program.

This path should take advantage of certain shortcuts to minimize disc access time and to relate more easily to the program's logic. For instance, place the master index frame near the center of the disc to minimize search time in both directions once the menu choice has been made.

Sequence placement at either the beginning or end of the disc is also a factor. Remember, the videodisc plays from the center to the outside edge. Avoid placing any "beauty" shots or sequences requiring high resolution at the disc center on a CAV (constant angular velocity) disc. Because data is packed on these tracks due to their short length, image quality is poorest. Also, don't hang important, frequently accessed still frames on the outside edge, because search time is slowest in this data wasteland.

A myth persists that you cannot mix stills and motion sequences in close proximity. It is actually much better design to place stills close to any related motion sequences. Proximity is an important consideration for menu choices as well. Those choices you feel will get the most play should have related still frames placed close to the menu for speedy search. A word of review: laying down more than one discrete still frame of a subject allows a larger search "landing pad," at the expense of a bit of disc real estate. Go for three to four frames, unless you are *really* cramped for space.

Finally, frequent fades to black between sequences solve three editing problems. First, they allow sequences to be edited out of continuity and laid off for later insertion into the master program. This allows editing to proceed even if all the sequences have not been approved, or produced. Second, in Level 2 programs, if fades to black are established as part of the program's style, then their occurrence will mask your 150 frame blackouts for digital dumps. Third, if for any reason you have to go back into a program and make a fix, the editor can go into the program at the nearest black fade, avoiding a complete re-edit. This advice also applies to cuts. While dissolves between scenes are beautiful, be sure to sprinkle in some cuts. If you have to re-edit, the editor can begin at a cut-point instead of having to make a match-frame with the dissolve, which can take time and cost money.

An interesting bit of knowledge is timely right here. We have been talking about placing your frames in specific places. When giving instructions as to placement of picture stops and chapter intervals, time code numbers are requested to mark the exact locations of these points. What if you want to be able to translate the time code number into an actual frame number? You can do this if three possible criteria have been met:

1. All of the original programming has been shot on videotape.
2. If film is the medium, then it has been transferred to videotape at 30 frames per second.
3. Film has been transferred using the 3-2 pulldown method and freeze-frame capability is not necessary.

A formula exists that makes this translation a short pencil and paper exercise.

First, imagine you have an edit point at time code 00:28:14:11 and you want to know the frame number of that point.

You must first convert the time code you are starting with as disc frame 0 into actual total video frames. We will call disc frame 0 the following:

00:00:05:05

This works out to:

00 hours	x	108,000 =	0
00 minutes	x	1,800 =	0
05 seconds	x	30 =	150
5 frames	x	1 =	5
			155 video frames

(Starting out with 5 frames and 5 seconds is arbitrary and not required.)

We then have to compute the actual frames up to the time code frame of our edit point:

$$
\begin{array}{lrrr}
00:28:14:11 & & & \\
00 \text{ hours} & \text{x} & 108,000 = & 00 \\
28 \text{ minutes} & \text{x} & 1,800 = & 50,400 \\
14 \text{ seconds} & \text{x} & 30 = & 420 \\
11 \text{ frames} & \text{x} & 1 = & \underline{11}
\end{array}
$$

50,831 video frames

To arrive at the specific frame number, we subtract the frame 0 number from the edit point number:

Videotape frames at 00:28:14:11   50,831
Videotape frames at 00:00:05:05   – 155

Videodisc frame number   =   50,676

When sending in your production order to the company that will do the actual encoding, be sure to indicate your frame 0 time code if you plan to use specific numbered still frames. Most disc duplication houses will request it in any case.

At this point, I could lay down some cost guidelines concerning video production houses, crews, editing sessions and equipment rentals, but they would not mean very much. Costs fluctuate according to locale. A shoot I budget in Chicago costs 30% more than a shoot in Fort Worth, TX, using the same equipment. This also holds true in the reverse for California or New York where prices are higher than in the Midwest. Giving price guidelines is not as valid as laying down some time benchmarks for an average 30-minute program. Using the following time schedule, you can work out your costs depending on your own locale and situation:

- The basic outline indicating content and program objectives: 15 to 20 days.
- A detail treatment of that outline showing the basic flowchart, script scenario and program strategies: 15 days.
- Treatment review time: 2 to 10 days.
- Revisions of that treatment: 3 to 5 days.
- Draft scripts (for example):
    100 still frames
    35 pages of narration (voice-over)
    10 page flowchart
  Total time: 15 to 20 days.
- Script/flowchart review: 3 days.
- Script/flowchart revisions: 3 to 5 days.
- Simple studio shoot with one set and one camera: 4 days.
- Offline edit: 2 to 5 days.
- Online edit: 3 to 5 days.

These numbers relate to an experienced production designer and show both best- and worst-case situations. As you can see, production time for an interactive videodisc is considerable. This time does not reflect digital programming time, which is necessary for Level 3 productions.

CONCLUSION

The design and production of your interactive video program is where the action is. It's the stage at which the *idea* of interactive video truly confronts *reality*.

The success of any video production relies on planning, but this is even more true with interactive programs, because there is so *much* planning. You are not dealing with a medium that elicits a passsive response, but one that asks the viewer to become an active participant—to respond to and become part of the program's design. This kind of manipulation carries a heady responsibility. In no other medium is the designer required to be so intensely attuned to the needs of the audience. Achieving this synergy is not easy, but it is very rewarding.

Planning and organization must begin with a clear idea of what the program's goals are to be. To accomplish this the task at hand should be approached analytically. Simply, you must ask, "what do we want to teach—or sell?" Is the solution to the problem really an interactive program—or could it be solved better with employee motivation? These are the kinds of questions you must ask. Once you establish the need, the method to achieve the teaching goal follows. In a sales situation—a point-of-sale kiosk for instance—you would need to explain the features and benefits of your product to a busy shopper in the most direct and involving way. There are many paths to good interactive design and they all begin with a good plan and good execution.

First you need a document—many documents actually—that thoroughly outline the task and parcel out the responsibility for getting the job done. This clear road map that traces the path from the beginning of the project to the end includes a flow chart. Once again—*many* flow charts. These charts are drawn on at least three levels: to help the client understand the flow of the program; to help the designer stay on track and control the branching; and to help the programmer keep track of the code needed for the structure.

With the plan completed, the execution follows. Actual program production is the realization of the plan and a test of its effectiveness. It's where the interactive program producer really earns his salt.

Finally, interactive programs are not cheap, due to their complex structure. Careful planning can save money in shooting, editing and computer programming. Budget dictates the realities of execution and level of expectation. A carefully managed budget, one that balances talent and hardware, shooting and editing schedules, and realistic time frames with availability of people and products, will extract the maximum screen value for the effort.

A successful interactive program is the perfect blend of design and production skills.

8 Hardware and Software Evaluations

There are, essentially, two ways to approach the creation of an interactive program. The first is to decide on a programming language, or authoring system that best fits the needs of the creative process, and then assemble the hardware components that will make use of the language. Many manufacturers of hardware components—circuit boards that fit into computers, stand-alone interface devices, user input systems, etc.—create their own authoring languages to tie all the components together. Some of these authoring languages are very complex and offer graphic overlay routines: commands to output to or receive input from keyboards, mice, touchscreens and light pens. Some even provide communications sections to allow porting information over telephone lines to other computers. Other languages are much simpler and allow the designer to work within the limited boundaries of the component system.

The other approach is to acquire a "turnkey," or "integrated," system. Such systems are sold as complete workstations, either of components produced only by one manufacturer—such as the Sony View system—or of a mix of proprietary components and other manufacturers' components. IBM's InfoWindow and the NCR InteracTV systems are examples of the latter.

We will survey some of the integrated systems and compare hands-on efforts with two authoring systems—one simple and one more sophisticated.

The turnkey, integrated systems offer one-stop shopping and one-telephone-call service, so we will begin there.

THE SONY VIEW SYSTEM

As a veteran of the early Microcomputer Wars (1978 to 1984), I learned one basic tenet: a system assembled from a variety of manufacturers' parts will present more difficulties than a turnkey system from a single manufacturer. That same rule holds true today for interactive video design systems. Sony was the first to assemble such a single-manufacturer workstation when it combined its SMC-70 microcomputer with the LDP-1000 videodisc player by means of a system of bolt-on interface controllers and a "smart" Sony video monitor. The system I used back in 1984 was impressive at the time, and many are still out in the market creating programs today.

103

We will look at this system briefly as a background for the evolution to the present Sony View workstation.

The component combination I used was comprised of the following:

- SMC-70 microcomputer
- Sony LDP-1000A videodisc player
- Dual floppy disk (3 1/2-inch) component bolted to the top of the SMC-70
- SMI-7073 Superimposer component bolted to the back of the SMC-70
- Sony PVM-1271Q 12-inch RGB/NTSC monitor
- Sony software:
  Sony Disc BASIC language
  Video Utility program for the Disc BASIC
  Video Titler
  Graphics Editor

With this system, I could search a videodisc for single frames or action sequences and overlay either text or graphics on the video image. I could also create text and graphics on color backgrounds to act as menus, or information "pages." To run an interactive program, the user only needs to insert the correct videodisc in the LDP-1000 player and the matching floppy disk in the SMC-70 Microcomputer, power up both machines and interact with what appears on the screen.

The SMC-70 Microcomputer runs the granddaddy of computer languages: CP/M (control program for microprocessors) from which the basic structure of today's MS-DOS was derived. Its RAM (random access memory) is 64K and it has the distinction of being the first microcomputer to use the 3 1/2-inch, hard-case floppy disk (a design driven by Sony). The SMC-70 is driven by a 10-year-old Z-80A micro-processor (still used in some videodisc players today) with a clock speed (regulates the speed of accepting and executing instructions) of 4.0 MHz (compared with the 8 to 20 MHz speeds of today's microcomputers).

Bolted onto the back of the SMC-70 was the SMI-7073 Superimposer. This device has its own input/output ports and permits the superimposition of RGB graphics and computer text over an NTSC video image. Another version—the SMI-7074—permits the superimposed RGB-NTSC combination to be converted to a single NTSC signal and recorded on videotape.

Combining the above components with the Sony PVM monitor, which accepts and synchronizes both RGB and NTSC signals, gives the designer a considerable variety of display possibilities.

The four software programs provided the programming muscle that gave the system its flexibility.

Disc BASIC was Sony's videodisc version of the BASIC language. It allowed the programmer to speak directly to the LDP-1000 videodisc player with commands such as:

- LDPSTART—Start the disc player's turntable.
- LDPSEG—Call up a particular video segment.
- LDPFRM—Find a particular video frame.
- LDPWAIT—Hold on a frame until another instruction is sent.

The *Video Titler* contains six type fonts in three different sizes, which can be filled with up to 16 colors and trimmed with edging or drop shadows. The *Graphics Editor* allows you to draw either thick or thin lines with a light pen, digitized drawing pad or the keyboard. Anything created with these programs can be superimposed over either a single frame of video or an action sequence.

The SMC-70 was a breakthrough, but learning its idiosyncrasies was a tedious process. The Disc BASIC language was very flexible, but mastering its syntax required weeks of study combined with trial and error tests.

The Sony SMC-2000

The Sony SMC-2000 computer and its mate the LDP-2000 videodisc player were combined in 1985 to create a quantum leap forward in workstations. (See Figure 8.1.)

This first Sony View System weds the various components, using inserted circuit boards to create an attractive package with a small desktop footprint. It also uses the industry standard MS-DOS operating system, which can make use of up to 640K of memory (compared to the 64K of CP/M). The floppy disks can accept up to 720K of formatted information. The microprocessor is the Intel 80186, a 16-bit (the Z-80 is an 8-bit) device operating at a clock speed of 6 MHz.

The SMC-2000 is an accomplished high-resolution graphics performer. Up to 256K of video RAM is available for high-resolution color graphic display and 16K for standard text and monochrome graphics. It is capable of offering 640 x 400 pixels of resolution, and each pixel can accept any of 256 colors chosen from a palette of up to 4096 possible hues.

For the designer, this means high-quality graphics superimposed over video images. These graphics can be overlaid in two planes: CG (character/graphic) and HG (high-resolution graphic) with the ability of adding transparent colors to either plane for a true two-plane superimposition over the video image. The CG and HG supers are RGB, and the video is NTSC. With the SMC-2000, the RGB and NTSC signals can be *combined* into a single, high-quality RGB signal. Using a Sony (or any monitor with RGB inputs) RGB monitor allows the best possible combined video-graphics image.

If you wish to videotape the result, you must go to a digital (RGB) to analog (NTSC) convertor, or make use of the latest Sony DVTR (digital videotape recorder).

Accompanying the SMC-2000 microcomputer is the LDP-2000 videodisc player, a front-loading machine that stacks above the computer and under the monitor to form a cosmetically pleasing package. The LDP-2000 offers a fast-frame search time of 1.5 seconds from frame number one to frame 54,000.

Actually, the LDP-2000 is a series of five machines that can be upgraded by the simple insertion of a circuit board. The LDP/1 has an RS-232 computer interface plug and baud rates (data sending rate in bits per second) of 1200, 2400, 4800 and 9600. The LDP/2 adds the ability to play discs with expanded command sets for Level 2 interactive playback. The LDP/3 adds an IEEE 488 parallel (RS-232 is serial)

Figure 8.1:   The first Sony View System.

Photo courtesy of Sony Corp.

computer interface permitting maximum data sending rates and controls up to 15 videodisc players from a single microcomputer. The LDP/4 permits the use of compressed digital audio. The ultimate configuration is the LDP/5, which combines all the features of the above together with the ability to read digital data—up to 220 megabytes—directly from the videodisc. Some academics refer to this direct data capability as "Level 4" interactivity.

Sony also created a more advanced programming language called BASIC/1. Also new was the *Graphics Title Editor,* an icon-driven drawing and titling utility that accepts input from keyboard, mouse or digital drawing tablet.

The Sony SMC-3000V

The most recent configuration of the Sony View System is the SMC-3000V microcomputer teamed with the LDP-2000 player. (See Figure 8.2.)

This computer is fully compatible with the IBM PC/AT standard, using an Intel 80286 microprocessor with a clock time of 8 MHz, and comes with 640K of user RAM. This standardization allows the SMC-3000V to use any IBM software for applications outside of its interactive design capabilities. There are five 16-bit expansion slots and storage capacity for a 40-megabyte hard disk or 1.4-megabyte floppy disks.

The graphics display capabilities have also been enhanced with a raise in resolution to 672 x 496 pixels and up to 512K of graphics RAM. This allows one graphic

Figure 8.2: Advanced Sony View System.

Photo courtesy of Sony Corp.

image to be displayed as another is being loaded from the hard disk. The result is a capability to flip from one graphics overlay to another, virtually instantaneously.

The Sony 3015 and Peripherals

For the first time, Sony has also offered a low-end companion to the fully configured View System. The Model 3015 combines the SMC-3000V with the new LDP-1500 videodisc player. The LDP-1500 is also front-loading, offers a 2.5 second search time and an RS-232 serial computer interface. Essentially, it provides basic playback capability without the expandable features of the upscale LDP-2000.

To round out these systems, Sony provides a copious list of peripheral options including a light pen, mouse, graphics tablet, touchscreen and both 12-inch and 19-inch PVM monitors. The software has once again been upgraded to match the new microprocessor/graphics capabilities.

The success of the Sony View System is reflected in the amount of pre-configured courseware available for it through a number of independent developers. A reasonably up-to-date listing of this courseware is available from Sony.*

IBM InfoWindow

IBM has returned to an earlier, less publicized, time when the corporation flirted with videodiscs. Back in the late 1970s, IBM spent considerable research and development money on the creation of a "smart" videodisc player based on its considerable computer expertise. At that time, however, the videodisc marketplace was anything but stable (see Chapter 1). In the end, IBM threw in its cards and pursued what it knew best—designing computers—and four years later produced the IBM PC. Now they are back in the videodisc arena.

In June of 1986, IBM announced a Level 3 videodisc interface and touchscreen input system called InfoWindow. It also introduced a collection of authoring and presentation tools and a system of pre-written courseware. InfoWindow has become such a success that many independent courseware and custom interactive system developers are writing their programs to run on the IBM architecture.

InfoWindow is the result of years of studies in interactive programming and hardware by IBM designers. Numerous prototypes and pilot programs helped determine what worked and what did not. The end product is deceptively simple in appearance and operation.

The hardware consists of an IBM PC and a color monitor. The monitor is the key to the system's effectiveness. Using piezoelectric force transducers, which create up to 60 "touch-points" on the screen, IBM avoided the more common, plastic pressure-sensitive membrane to achieve a brilliant enhanced graphics (EGA) image. This InfoWindow display is actually a "smart" terminal complete with its

---

*For information write to: Sony Intelligent Systems, Sony Corporation of America, Sony Drive, Park Ridge, NJ 07656.

own microprocessor. It permits the overlay of RGB graphics over NTSC video within a range of 16 screen colors from an available palette of 64 hues. Text can be added in 40 or 80 column formats, producing up to 25 lines per screen "page" over color backgrounds. Measuring 13-inches diagonally, the display sits atop any IBM personal computer from the original PC through the XT, AT and the new PS/2 generation. Since many training and office locations already have these machines, interactive program creation can be implemented with a minimum equipment investment.

The following Pioneer videodisc players can be used with InfoWindow:

- Pioneer LD-V6000 (with or without the SWSD Processor; SS-D1 for sound over still frames)
- Pioneer LD-V6200 (two LD-V6200s can be connected to the InfoWindow display to increase program presentation flexibility).

Other features include a voice synthesis chip with a pre-defined vocabulary, dual speakers with the ability to support stereo sound from a disc's dual tracks and video projection to a large screen.

The elegant simplicity of the hardware configuration is mirrored in the authoring and presentation system: IBM Learning System/1 (LS/1). This program requires no previous computer programming experience. It attempts to bridge the gap between the flexibility of an authoring "language," such as Sony's BASIC/1 (and the attendant complexity) and the structured (but often limited) menu-driven authoring "system." Using a PC mouse, the designer can select from a very comprehensive menu. Training programs can include the following capabilities:

- Determine correct and incorrect answers.
- Create "student help" information to accompany the question.
- Set answer time limits.
- Allow for typing and space errors in student answers.
- Determine whether a student should continue the program with a correct answer or receive additional help for an incorrect answer.

A set of "Presentation Aids" is also available. They include:

- Video Aid. Co-ordinates frames of videodisc images with computer-generated information.
- Graphics Aid. Creates new graphics or makes use of graphics created with another IBM program.
- Window Aid. Creates pop-up screen "windows," or text screens, that can contain material from other files.
- Speech Aid. Makes use of the speech synthesis chip and its built-in vocabulary to give vocal instructions.
- Multiple Authors. Allows several authors to work on the same program. For instance, a team of authors can work on the lesson's presentation while another team concentrates on the educational content. Their combined efforts can be merged into the final program.

- Presentation Aids Controller. Allows editing and sequence changes of presentation features.

Finally, the program can be checked at any time with the "Run" feature, so the program can be viewed as the student will see it.

To run this software, the IBM personal computer (or a very compatible work-alike) makes use of the following configuration hardware as requirements vary:

- 640K of RAM for the Authoring System
- 512K for the Presentation System
- Minimum 10MB hard disk drive
- One 360K 5.25-inch disk drive
- IBM Enhanced Graphics Adapter (EGA) with 128K additional RAM
- IBM EGA Jumper Card
- IBM General Purpose Interface Bus Card and cable
- IBM Asynchronous Communication Adapter (or functional equivalent)
- IBM InfoWindow Display
- Mouse Systems™ PC Mouse (or functional equivalent). Required for Authoring System; optional for Presentation
- Videodisc player (Pioneer LD-V6000 or LD-V6200).

The following software is also required:

- IBM Disk Operating System (DOS) Version 3.20
- IBM InfoWindow Control Program.

To access the considerable volume of courseware created for the InfoWindow system, IBM has created the Market Assistance Program (MAPS). This program gives users access to a network of qualified vendors and designers of compatible InfoWindow courseware, which can be customized to specific needs either on a consulting or complete turnkey basis. To date, over 110 complete InfoWindow courses have been identified.

NCR InteracTV-2

This system is based around the NCR PC6 Microcomputer, which is actually an IBM PC work-alike based on the MS-DOS operating system. It uses the Intel 8088-2 microprocessor, which gives a switchable clock speed of 4.77 MHz or 8 MHz. What this means is that if you already have an IBM PC/XT, or compatible machine, all you need are the NCR video and graphics circuit boards to begin building the system.

The video controller board, which fits into a slot in the computer, provides all the necessary graphics and monitor functions. The board is fully compatible with the standard PC color graphics card and provides an enhanced display of 640 x 400 pixels (standard display is 320 x 200 pixels) with a palette of 16 colors. This card contains:

- 128K of RAM.
- 64K of EPROM (Erasable Programmable Read Only Memory), a microchip programmed at the factory and capable of being reprogrammed, or updated. It can only be read and not written to by the user. Unlike RAM, the information stays in the EPROM when the power is switched off.
- A 6845 CRT (Cathode Ray Tube) controller chip.
- RGB output.
- Horizontal and vertical sync outputs (controls image scan).

The video controller also takes up one slot worth of space on the computer's motherboard. This card provides all the inputs and outputs.

A pair of audio speakers is also offered. They plug into the rear of the video controller card, which protrudes out the back of the computer chassis. Each speaker can handle up to 50 watts with a 4 ohm impedance. Either stereo sound or two language tracks can be handled by the system.

Originally, the InteracTV system made use of the Pioneer LD-V1000 Level 3 videodisc player. That machine is no longer manufactured, but the current LD-V6000 player series works with the InteracTV-2. The Hitachi VIP-9550 and Phillips VP 831 (PAL standard) are also compatible. Compatibility is determined by the software drivers offered by NCR as well as by the need for an RS-232 or parallel computer connector plug.

Software for the system includes what NCR calls "Toolware." These utilities come on a floppy disk and provide the user with configuration programs to define the hardware environment (disc players, printers, any input/output devices), drivers to allow communication between the computer programming language and the input/output systems, touchscreen calibration and video system diagnostics.

Writing interactive programs for the InteracTV-2 can involve general purpose languages, such as BASIC which links to the videodisc player via the "Toolware" or authoring languages and systems. A list of authoring programs is available from the manufacturer.

InteracTV-2 Paintbrush is graphics editor software that allows you to create on-screen RGB graphics. You can use the necessary video mixing, videodisc player control, video overlay, touchscreen control and audio amplification. Essentially, it accepts the NTSC color video signals and mixes them with the RGB graphics signals and ports them to the display monitor. It contains:

- A videodisc player input/output controller.
- The monitor's touchscreen interface.
- A two-channel audio amplifier.

The standard InteracTV-2 monitor is a 14-inch, analog RGB display operating at a high-resolution bandwidth of 24 MHz. It can accept either NTSC or PAL (Phase Alternating Lines—a European video standard). This monitor is capable of excellent imaging, but it comes into its own when combined with the optional NCR touchscreen.

This screen responds to the touch of the user by means of a transparent, electrically resistant membrane attached to the monitor's front. The screen is connected to the video controller card, which provides a bias voltage to the membrane components. The user's finger compresses the membrane by .001 to .005 inches. The computer samples the touch point, digitizes it, then sends the X and Y coordinates to the processor. Two to eight ounces of pressure are required to trigger the screen. A stylus similar to a ball point pen, which exerts a pressure of one to three ounces, can also be used. Conversion time from touch to response can be as fast as 50 X/Y coordinates per second, or 20 milliseconds per point. These X/Y values are converted into parameters that your software program is waiting to receive in order to send a command. Touch resistant screens are the most common interactive method for input. The more traditional keyboard, mouse or digital graphics pad can also be used. Touchscreens tend to "dull" the screen image brilliance due to the nature of the plastic membranes, but NCR's enhanced graphics and high-resolution bandwidth overcome this appearance of "fogging."

The NCR InteracTV-2 is a no-nonsense system based around proven technology. Its screen images are made up of RGB graphics mixed with NTSC-converted-to-RGB video to produce sharp, brilliant pictures.

INTEGRATED SYSTEM SURVEY

While the major integrated system market players are IBM InfoWindow, Sony View, and NCR InteracTV-2, there are other packages that claim a fair share in their own market niches.

The Matrox LVC-2001AT

The Matrox LVC-2001AT is the core unit in the Army's new EIDS program (see Chapter 5), which is comprised of a massive network of military training workstations. It uses an IBM AT working off an 80286 16-bit microprocessor running at a credible speed of 10 MHz. The basic RAM configuration is 512K and an Enhanced Graphics Adaptor (EGA) card is responsible for text and graphics overlays with a resolution of 720 x 480 pixels.

The disc player for the Matrox system is the Hitachi VIP-9550, which is capable of reading data storage off the videodisc and providing sound over still frames. The display is a standard 13-inch RGB or NTSC monitor.

The ITS 3100

The ITS 3100 is offered by Interactive Training Systems. It is based around an IBM PC with 340K of RAM. An interactive board set controls the videodisc player and provides graphic overlays of 640 x 400 pixels resolution. A switchable monitor capable of either RGB or NTSC completes the package. ITS has its own authoring program called *Authority* or uses Allen Communications *Quest* software.

Online Computer Systems

Online Computer Systems has created a system based on its GL-512 overlay and videodisc controller card with genlock (NTSC video synchronization) capability. This system is a regular mix-and-match blend of manufacturers' components. It can be based around an IBM PC, XT or AT, uses either a Sony LDP-2000/1 or Pioneer LD-V6000 videodisc player and feeds the signals into a Sony 1271Q monitor with touchscreen. Digital voice output is an option as is the ability to network between other computers via modem.

The Pioneer LD/VS-1

Pioneer offers its own system, the LD/VS-1. This package is built around an MSX-2 microcomputer using a microprocessor equivalent to the venerable 8-bit Z-80A—a survivor from the 1970s. This major component is called the UC-V102 LaserDisc Controller. It has three expansion slots, a 3.5-inch floppy disk drive capable of storing 720 Kilobytes and a ROM cartridge expansion plug typical of the MSX computers in use in Japan. (The MSX system never found a real home in the United States.)

The LD/VS-1 has a single RS-232 videodisc interface, but a system of multiplexors allows control of up to five players. A graphics overlay is possible, and the video circuitry outputs either RGB or NTSC signals. Its complex switching configurations allow computer generated sound to be mixed with videodisc audio tracks, and input can be selected from as many as three different composite (RF) audio-visual systems.

The Pioneer LD-V4200 is the player offered with the basic "A" system together with a TVM-V1300 monitor capable of RGB or NTSC display and a UK-V101 keyboard. Pioneer also supplies an optional UK-V105TS resistive touchscreen.

Learncom/Videologic

Learncom/Videologic builds the MIC-2000 and MIC-3000 systems. The MIC-3000 works off an IBM PC or compatible using an IVA-3000 video controller/overlay circuit board. It provides EGA graphics overlay and the flexibility to operate any videodisc player through its RS-232 interface port.

MIC software makes use of MIC-DOS, an extention of MS-DOS that provides interactive video commands. When combined with the MIC Virtual Device Interface software, the user can control videodisc players. Another feature allows a version of the software to be used on an IBM InfoWindow, permitting porting of programs from one system to the other.

The Visage V:Station 2080

Visage offers the V:Station 2080, an IBM AT-based system that supports most videodisc players. Its graphics capabilities come from a Texas Instruments T19128 graphics processor permitting a 16 color overlay of 256 x 192 pixels resolution. The V:Station 2080 supports a considerable catalog of available authoring languages, including: *Quest, CDS, Genesys, Educator, IMSATT, SAM, Video Nova* and *Tencore*.

The Comsell Prism IV

Comsell has flourished in the realty market and offers the Prism IV, a system built around the Intel 8086 microprocessor. It provides IBM CGA graphics, RGB graphics and video signal processing. It supports any RS-232C-controllable videodisc player. A unique feature is the IV Gold Card—a ROM circuit the size of a credit card that carries 256K of RAM—for insertion of either operating systems or courseware developed by Comsell.

The Integrated System Marketplace

Integrated system hardware configurations proliferate in the interactive market-place—each offering its own spin on the ball. It is a market not unlike the Computer Wars of the late 1970s to early 1980s when a number of proprietary solutions scrambled for market acceptance and a share of user loyalty. Only the IBM computer standard seems to be acknowledged, but there are also systems built around Apple Macintosh, Digital's IVIS and the Apple II computer family which has deep educational market penetration. It is difficult to base long-range purchase decisions on which standard will survive.

COMPONENT INTERFACE SYSTEMS

An alternative to the integrated "turnkey" system is a collection of components, the most important of which is the connection (interface) between the computer and the videodisc player. This piece of hardware usually comes with its own software command set on a floppy disk. If you already have the necessary microcomputer, this approach is obviously less expensive, but there are trade-offs. Any compilation of components assembled from different manufacturers carries its own caveat of "implied compatibility." This means YOU are responsible for making all the right hardware hook-ups, threading cables and snapping together adaptors. YOU are responsible for meeting each manufacturer's requirements, flipping dip-switches, plugging and unplugging circuit boards. YOU must make the phone calls to the various manufacturer service reps to beg for information when the system hiccups.

Still, component systems are an alternative to the more expensive integrated systems, so to study the subject first hand, I ordered two systems—one for the Apple IIe I owned at the time, and one for an IBM PC I had on loan for the test. The two packages differed in sophistication. They also represented opposite ends of the spectrum in cost, hardware and their interpretation of the interactive spirit. They are included in this book as examples of their kind rather than as critiques of the products themselves. These are learning experiences—diaries of discovery—examining what can be expected in the world of component interface systems.

USING THE APPLE IIe

The Video Vision Associates Authoring System

The Video Vision Associates VAI II videodisc player interface is a small, white box about 1 1/2-inches high by about 3-inches wide and 2-inches deep. It looks like one of those splitters you buy for your TV antenna so you can feed more than one receiver. (See Figure 8.3.)

The VAI II comes with an authoring language called *Laser Write*. The combination allows you to create an interactive program using existing images on a disc and two screens. You create what is called a "Playback" disc, which holds the program you create with *Laser Write*. When the program is run, you are aware only of the program and not the language involved. Video Vision calls this procedure a "shell."

Laser Write requires no programming experience. A list of single and two-keystroke commands is provided that allows you full control over the disc player. This type of program is not as flexible as a true programming language such as BASIC/1, but *Laser Write* is designed to appeal to nontechnical users who are not

Figure 8.3:   The Video Vision Associates VAI II Videodisc Player Interface.

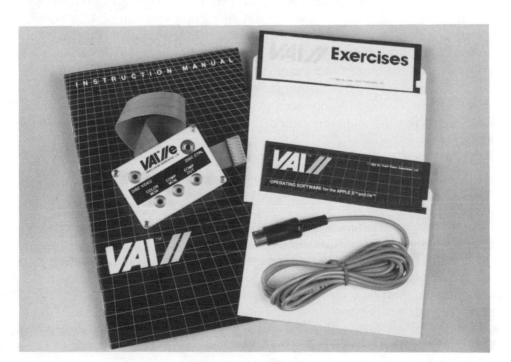

Photo courtesy Video Vision Associates.

interested in sophistication, but have a need for Level 3 interactivity. *Laser Write* also can be used in computer-aided instruction (CAI) without the videodisc control option provided by the VAI II. This CAI format makes use of the computer screen only. The VAI II is interesting because of its ability to switch between video output from the videodisc and text from the computer to the color monitor. Computer text remains on the computer monitor at all times during the program.

Since the hardware is somewhat unimposing, software is called upon to provide all the muscle. The VAI II Apple Interface attaches to the Apple joystick port in the right rear corner of the cabinet. This type of connection illustrates the usefulness of a game port splitter such as the Kraft E-Z Port II. This device allows you to attach two peripherals to the single port via a cabled extension outside of the cabinet. Zero-force connectors allow easy insertion of those terribly fragile, tiny pins. The combination of simple hardware and authoring system blends in a comparably inexpensive package.

The system offers a comprehensive set of commands for the Pioneer LD-700 and LD-V4000 players. I have the LD-V4000 machine and the interface connects to the remote control, 8-pin connector on the front of the unit. An alternative connector is also included. This black half-dome covers the machine's remote sensor that reads infrared signals from a wireless player remote control. A small, internal lamp in the dome takes the place of the remote's transmitter and connects to the VAI II interface via a mini-plug. Using both connectors allows you to use the player computer controlled or remote controlled without switching connectors around. The infrared dome must be used with all players except the LD-700 and the LD-V4000.

The VAI II box has five connectors: one RCA plug connects from the interface computer monitor jack to your computer monitor's video-in jack. Another RCA plug connects the interface computer out with the Apple video out jack. Then you hook an RCA plug from the VAI II color monitor jack to the color monitor's video-in jack. A final connection requires an "RF" connector attached from the VAI II box to the disc player's video output port. On the LD-V4000, this port is a BNC (standard video component) connector, so you have to buy an adaptor from BNC to RF. Except for the audio connections, that's about all there is to hooking up the little box.

Laser Write Software

Laser Write is written in Apple ProDOS, and the VAI II instruction manual takes great pains to make sure the user understands the ProDOS environment and how it pertains to the disc operating structure. Though ProDOS is largely transparent to the casual user, its COPY A VOLUME command is used to back up the VAI II master operating software disk. Once this backup procedure is finished, the VAI II menu is available on the backup disk. For our purposes we will choose selection 5, PLAYER SETUP. A list of players is presented, and you select the one you intend to use. The VAI disk will be configured for that player. (Be sure you do not write-protect your backup disk with tape over its envelope notch.)

   In order to check your connections and configuration, select option 3 from the menu, either TUTORIAL-MONITOR or TUTORIAL-TV, depending on whether you are using a video monitor or television set. The tutorial will present an example of a disk control program, and you can use any CAV videodisc as a control example.

   Video Vision Associates has created a series of videodiscs under its own label: the *Space Disc* series. These discs take advantage of the vast store of NASA photography, combined to create a collection of true to life adventures with titles such as: "Voyager," "Apollo," "Shuttle," "The Sun" and a very comprehensive production— "Astronomy." The "Apollo" disc tells the story of Apollo 17 and its voyage to the moon. The tutorial is designed to walk you through a videodisc presentation of highlights from this program. If your configuration and interface are properly set up, you will arrive at the proper locations on the "Apollo" disc as you follow the tutorial commands.

   Next, you can select the floppy disk entitled "Exercises" and run through a series of learning exercises to help you understand the VAI II commands. A reference card is also included with the literature that presents all the videodisc control commands. For the LD-V4000 Pioneer player, these commands are:

N	Step forward
F	Frame search
S	Search
P	Time a video clip
Q	Clear search command
L	Left audio only
T	Left and right audio on
X	Slow play forward
>	Multi up of slow play
I	3X play forward
H	Scan forward
Y	Pause
<	Computer video to monitor
B	Step backward
C	Chapter search
W	Wait for completion of search
Pn	Timed playback of n time units
·	One second delay
R	Right audio only
D	Chapter and frame display
Z	Slow play backward
(	Multi down of slow play
U	3X play backward
G	Scan backward
&	Reject
)	Videodisc video to monitor
#	Repeat "A" command

∂2   Laserdisc video off
∂4   Laserdisc audio off
∂6   Laserdisc character generator off
$ Repeat "B" command
∂3   Laserdisc video normal
∂5   Laserdisc audio normal
∂7   Character generator on

These keystrokes are linked in strings to produce a complete command, which also includes a disc address. For example, the command to search frame 1000 would be SF1000SW. The "Exercises" floppy disk explains how these strings are used.

Besides the videodisc layer commands, another list of primary functions is also offered. These use single keys to create commands such as: A = create string of videodisc commands; and control-key commands such as: CTRL–L = load file from disc. Playback functions list strokes including: S = begin sequential playback at a selected string; or ? = list strings in memory.

With an understanding of these commands, you have a few alternatives open. First, you can write your own program by formatting a new disk using the ProDOS FILER utility (Option 7 on the VAI II menu). Then, you can use Option 6 to store the proper videodisc player driver on your new disk. After that, you load the SUBROUTINE file off the VAI II master disk and then list the SUBROUTINE. Listing a program shows all the line numbers and the program's structure. If you have used the string command to search frame 1000, you will see that the SUBROUTINE program ends at line number 1000. You can begin writing your program on the next line and continue until line number 6000 if you wish. When the writing is finished, you can copy the new program onto your new disk.

The *Laser Write* Manual

In order to grasp the full potential of this reasonably versatile authoring system, stick with the instruction manual that accompanies your *Laser Write* software. The literature promises that, after about three hours of application, you will be familiar enough with the program to begin experimenting on your own. Of all the manuals I have worked with over the course of writing two books on computers, this 15 page introduction to *Laser Write* is one of the clearest I have ever seen.

The manual is broken up into chapters. The first acquaints you with the basic principles and walks you through the exercise of making a duplicate copy of the *Laser Write* master disk. Chapters 2 through 4 provide hands-on tutorials and the opportunity to create your first interactive videodisc program. The final, sixth chapter offers advanced features of the system.

Laser Write is essentially an authoring program that allows non-programmers to develop a custom teaching aid through a question and answer prompted structure. The end product is a dedicated disk that runs your program and the *Laser Write* tools that aided the writing are transparent to the user. This end product is called the "Playback" disk.

Using *Laser Write*

To step through the process of creating the playback disk you have to under-stand some basic terms used throughout the manual's organization of playback data. These include:

- Page number (A "page" is one screen of text.)
- Page style
- Videodisc commands
- Text to be displayed
- Branching information

Creating Pages

Page style is broken down into three formats: caption, true/false question and multiple-choice question. At this point, you are actually dealing with a dual screen Level 2 program, but for the beginner or non-programmer its variations are ample. Branching information simply tells the playback program which page of informa-tion to run next. A multiple-choice question could take the user to a different page of text for each possible choice. The entire program is based on this question and answer prompting method which is necessary to construct each page of the pro-gram. The pages are numbered and stored in files on your playback disk. Branching can occur between pages within a file or between files, which allows for quite a bit of flexibility.

Using a copy of the *Laser Write* master disk, the first tutorial in the manual takes you through the process of initializing your particular videodisc player. This is called "Laserdisc Player Setup." In the manual it is followed by the "Apollo 17—Mission to Taurus Littrow" exercise. (In Chapter 4 of the user manual, you will modify this Apollo program as part of the tutorial.)

In the "Mission to Taurus Littrow" exercise, both the color monitor showing videodisc images and the interactive program on the computer monitor are exercised in a tutorial showing how the text blocks (up to 16 lines) and questions posed by the computer relate to the video images.

Once you become familiar with the program, you can set up the system to toggle back and forth between text and color video images on the color monitor screen. Since all text is limited to 16 lines of 37 characters, the color monitor with its limited text resolution capability can handle the alphanumeric information with no difficulty.

Page styles offer either true/false questions or multiple choices that branch the user to other text blocks and video images. A simple space bar tap will also move the user from text page to text page.

Text Entry

As we move through the tutorial, we come to a section both in the manual and within the program in which you are constantly asked if your selection from the offerings and typed inputs is correct. Text entry requires a bit of extra thought.

An additional character (∧) must be added to the first column of the next line after your last line of text has been entered. This represents the keystrokes SHIFT-6 on the Apple IIe and the SHIFT-N on the Apple II+ and tells the authoring program that text entry is complete. If your Apple II+ does not have a lower case chip installed, then all copy will be in caps. "Yes-No" prompts do not recognize lower case letters.

When entering text captions, the program assumes each consecutive page of information is sequential. To indicate an out-of-sequence page, simply type the branch page number. The same applies if you are branching from one file to another. Just type in the branch file name. While doing all this text typing, you have to understand that you are using a simple line writer, not an editor. If you make a mistake, you have to re-enter the EDIT mode to make a correction. Even the line editor instructions caution you that the line editor is not a word processor and has limitations. When you access the EDIT mode, you can select the page you want to edit, but to exit you must advance through all the subsequent pages and verify each page.

Blocks of Memory

Every time you wish to produce a program, a variety of files from the *Laser Write* library must be transferred to a ProDOS formatted disk. The program can be any length you wish, but no file you create can contain more than 38 pages. The creators of *Laser Write* claim that programs as long as 200 pages have been saved to a single disk (in several files) but recommend moderation. The authoring system offers a constant reminder of available memory through a display of ProDOS "Blocks Free." There are about 218 blocks, or 430 sectors, free for use in a new, formatted, *Laser Write* playback disk. As you write text pages and follow the question "styles," this number of blocks will diminish.

Evaluation of *Laser Write*

For users who want to start with a simple system, *Laser Write* can be used with the Apple alone in the Computer Aided Instruction mode. This may be very mundane when compared to using languages such as BASIC/1 and *Super Pilot* that offer greater flexibility and also provide graphics and math capabilities.

Laser Write and the VAI II Apple Interface combine to create a framework that wears two facades. On one side, the modeling is restricted to specific question types: yes/no and multiple-choice, which is not unlike straight Level 2 programming with a Pioneer or Sony disc player remote keypad. It also requires the user to follow strict formatting of text screens. There is no provision for polling students, storing response data, or other more sophisticated Level 3 attributes.

The other side of the framework is a breath of fresh air. The program is truly easy to learn. In less than 45 minutes from fingers first poised above the keyboard, you are programming up a storm. It seems to have been designed with the teacher or company trainer in mind. This program is written for a person who does not want to deal with complexity, since the teaching plate is usually rather full of other chores. Any CAV videodisc can be grist for this person's interactive mill. Flowcharting is the only prerequisite outside the program's immediate domain. Video Vision Associates even gives you data sheets to help you lay out text pages. Their approach guides the user every step of the way.

Once the basics have been mastered—and it is fun to push the program around a bit to see what happens—you can move on to more advanced applications of the structured format.

The VAI II Apple Interface and the *Laser Write* authoring structure have traded off flexibility for simplicity, but the trade gives value received for entry level videodisc interactivity.

USING THE IBM PC

The Whitney Interactive Video Authoring Language

Since the interactive industry is quite new, some manufacturers maintain a very small inventory and are unable to deliver merchandise promptly.

The Whitney system, which is designed to work on the IBM PC microcomputer, arrived only two days after it was promised. It included a loose-leaf manual, two program disks, plastic bags full of color-coded cables and a circuit board to fit in my IBM.

This was the first opportunity I had to examine the viscera of the IBM PC. Its cabinet does not allow easy access as does the Apple, but once I had managed to remove the entire outside shell (6 screws), I noted that the tidiness of the construction made it worth the struggle. The machine does not offer many slots for expansion since so many basic functions require add-on circuit boards. In my case, two of the five slots were occupied by a 256K memory expansion board and a color monitor board. The Whitney card could slip into any of the remaining empty slots. An RS 232C plug exited out through the rear of the machine at the end of a ribbon cable.

The Whitney circuit board offers a bank of RCA plugs to mix and match a number of peripheral possibilities. One thing must be understood at this point, however. The Whitney interactive system was originally designed for interactive *tape recorders.* This becomes apparent as the RCA plug ports are identified. (The videotape recorder and time code plugs do not interest us.)

Insight PC Plus

The dual service—tape and videodisc—is a factor with which you must contend when learning the *Insight PC Plus* software that is used with the IBM PC. However, let us assume the hook-up of boards and cables has been completed and we are now finding out what Whitney has served up for us to comprehend.

To begin with, the Whitney Interactive Video System is a Level 3 system. Unlike Video Vision Associates' *Laser Write,* however, Whitney's software takes advantage of the computer's ability not only to create text and branching, to control peripheral devices and to permit interactive dialogue, but it also records student scores and responses. In other words, the computer is not simply a delivery system for a fixed learning structure, it is an integrated partner in the evaluation process.

The *Insight PC Plus* language is the heart of this interaction, and its structure is based around a vocabulary of 56 words. A number of tutorials are available for the user to sharpen his/her understanding of this syntax. To begin with, the authors of the program—and the authors of most other true Level 3 interactive programs—make some general assumptions.

It is assumed that the user is a professional teacher or is training staff members and is familiar with instructional theories. Next, it is assumed that the user is familiar with audio/visual instruction materials and the attendant jargon and resources. Finally, a familiarity with the potentials and limitations of computer aided instruction (CAI) is taken for granted. The authors also assume you have *and are familiar with* the IBM PC and its PC-DOS (MS-DOS) operating system and attendant utilities. You must also have access to one of a list of videotape recorders—or in our case one of three possible videodisc players:

- Pioneer PR-8210
- Pioneer LD-V4000
- Pioneer LD-V6000

These videodisc players will be designated either Model 1, 2 or 3 in the program's syntax. I received the cabling for the LD-V4000, so I used Model 2 in my program.

The Whitney User Manual

With assumptions behind us, a metalanguage of seven key words is introduced up front in the Whitney User manual so we can see the basic structure of *Insight PC Plus*. I will quote directly from the manual:

> —COURSE—This is the largest frame of reference used. A COURSE is a complete series of lessons designed to convey a significant body of knowledge or skill to a student. "Learning to Read" might be the title of a COURSE.
>
> —LESSON—A series of UNITS designed to be experienced by the student at a single sitting. "Words that start with 'A'" might be the title of a LESSON.
>
> —UNIT—The basic group of computer instructions in Insight PC. All Insight PC instructions must be gathered into UNITS. Lessons are planned and diagramed on a flowchart in terms of UNITS. There are two kinds of UNITS: PRESENTATIONS and QUESTIONS. These two types cannot be combined into one UNIT, but must be constructed separately. Every UNIT must begin and end with a directive and is composed of commands.

—PRESENTATION—A UNIT which offers instructional material in the form of videotape [videodisc implied here, but not mentioned—author] sequences, computer graphics, computer text, or a combination of two, or more of these.

—QUESTION—A UNIT which evaluates student comprehension based on answers to either a multiple choice or match query AND which can alter the instructional sequence based on the answer received.

—COMMAND—An instruction to the computer to do something either with the VTR or with the graphics or text stored in its own memory. "Play to Position ###" is a COMMAND. A UNIT may contain no more than 15 COMMANDS.

—DIRECTIVE—Like the sides of a box, DIRECTIVES work together to contain the commands occurring within a UNIT. "PRESENTATION 3" is a DIRECTIVE, as is "QUESTION 1." "NEXT" is also a DIRECTIVE. A DIRECTIVE must always appear as the first and last statement in a UNIT.

This metalanguage gives you a pretty good idea of *Insight's* basic concepts. Next, the manual teaches you how to fashion a working *Insight* Lesson Delivery Disk.

This is where things became slightly difficult for me. I had used the CP/M operating system for about three years, so I was used to creating my working floppy disks by copying over utility files from the system master disk. MS-DOS is a CP/M work-alike in many respects. *Insight* asks you to boot your MS-DOS master disk in drive A with your *Insight* Lesson Delivery Master in Drive B. At the A> screen prompt, you type B; CREATE and your DOS system is wedded to the *Insight* Master. You then use the IBM's *Diskcopy* program to create a back-up disk of the Lesson Delivery Master. I accomplished this with no fuss.

At the bottom of the next page of the manual was the word "EDLIN." The user was told that, if he/she did not have a compatible word processing program, a special program built into the IBM PC operating system called *EDLIN* could be used. *EDLIN* is a derivitive of *ED,* which was created for CP/M users back in the late seventies.

EDLIN allows you to enter and edit lines of text. You call it up from the MS-DOS files by typing EDLIN opposite the A>, followed by a space, and then the name of your Lesson Description (Program Name in *Insight* terminology). If the program exists as a file on your Lesson Delivery Disk, then the words "End of file input" pop on the screen, followed by the number "1" and an asterisk. If this is a new file, then you read "New file," followed by the "1*." If you are creating a new file, type "1" for "Insert," hit the Enter key and the "1*" jumps over a few spaces, ready for you to type in a line of programming. When finished, punch Enter and "2*" appears. You progress on through the entry of your program as each line number is consecutively incremented by one. Mistakes can be edited by pressing the Break key (Break on the key's front and Scroll Lock on top). Then you retype the line the way it should be. Typing LIST verifies the changes. To return to the program at line number 4, type "4I" (Line 4 Insert). You can also write explanatory lines that will not have an affect on the program's execution if you precede the line with a "∂." *EDLIN* also allows you to append lines, delete lines and divide a program into chunks if it is too big for your machine's RAM.

The alternative to using *EDLIN* is to load up *Wordstar* or any other word processing program that produces standard ASCII text files. This program will allow you to make block moves, line and word deletes, format, word wrap-around—all the little touches that make writing fun. *Insight's* programs are created as simple text files.

When you have finished typing in all your lines of program, you hold down the CTRL key and punch Break. This causes the screen prompt to jump back over to the edge of the screen from your column of program lines and you type "E" for "End" of the program. The program is saved onto your Lesson Delivery Master disk in drive A. Making sure the *Insight PC Plus* Language disk is in drive B, you type "B:Build" after the A > screen prompt.

"Build" is the key to *Insight*. It appears to be a compiler of sorts. There are two kinds of languages—interpreted and compiled. Interpreted languages look at each statement and then interpret it into computer machine code. A compiled language lets you type out the whole program and then compiles it into machine code in a lump. If you have erred, then you must mend the error and have the program recompiled.

"Build" acts like a compiler. It strokes through your program lines and spits out the errors along with possible suggestions to remedy the mistake.

Evaluation of Insight PC Plus

One problem with *Insight* is its orientation as an interactive video*tape* program. All the language writing tutorials, examples and explanations are in the front of the manual while the video*disc* commands languish back in Chapter 12. I had to wade through many pages before I could find the writing information I could apply to the videodisc lesson.

Petty problems aside, *Insight PC Plus* is an elegant program that allows maximum results from minimum programming input. The commands are simple and in plain English. Figure 8.4 is a sample and is almost self explanatory.

If you examine the program carefully, you do not really need much explanation. The program is grouped into units, presentations of text and questions (choose A or B). "Model 2" indicates the LD-V4000 player. The disc player "seeks" videodisc frame 3500 while the computer program "waits," then the machine begins playing to videodisc frame 4500 as the "VTR" command switches the images to the color screen. The computer program "waits" for the play to finish. "Text page 2" appears on the screen as "IBM video" is switched on. This could overlay the text on the image if you are using a Sony-type monitor, which performs the overlay function, or toggle between the images and the text on a single screen. At the bottom of the program, the text for each PAGE is listed. The program dips down to these lines when choices are made by the user. For BASIC language programmers, this is the same as the "Read Data" function.

Figure 8.4:   *Insight PC Plus* **Commands**

```
PROGRAM 'PRACTICE.PRG'
MODEL 2
UNIT 1
PRESENTATION 1
ASSUME 'DISC'
   SCREEN 1
   TEXT PAGE 1
   DELAY 2
   SEEK POSITION 3500
   WAIT
   PLAY TO 4500
   VTR
   WAIT
NEXT
END OF UNIT 1
UNIT 2
QUESTION 1
   TEXT PAGE 2
   IBM VIDEO
   PROMPT
   IF —A—
   TEXT PAGE 3
   DELAY 4
   SHOW FROM 3750 to 4000
REPEAT 2
   IF —B—
   TEXT PAGE 4
   DELAY 3
NEXT
END UNIT 2
UNIT 3
PRESENTATION 3
SHOW FROM 5000 to 6500
NEXT
END UNIT 3
BEGIN PAGES
] Page 1
This is Page 1
^
] Page 2
What would you like, A or B?
^
] Page 3
You choose A.
^
] Page 4
You choose B
^
EOT !
```

There is also a set of direct videodisc player control commands. They include:

DISC 'SCAN FWD'
DISC 'SLOW FWD'
DISC 'STEP FWD'
DISC 'AUDIO1'
DISC 'FRAME'
DISC 'SCAN REV'
DISC 'SLOW REV'
DISC 'STEP REV'
DISC 'AUDIO2'
DISC 'PLAY'

These commands can be inserted in the program as though you were pressing buttons on the remote control.

Insight allows the use of IBM's color graphics to draw overlays and enhance the CAI screens. Lessons can be personalized by embedding the student's name in the text pages. For example: ":N$:", you have forgotten an important rule." Phil, Charlie, or Beth becomes ":N (Name) $ (Character string) :."

There is even a command that covers "creative" students who do not choose any of the offered choices. It's called **ON NOMATCH** and is added to the program whenever a **MATCH** is required in multiple choice or a character string response is necessary. **ON NOMATCH** simply boots the user back to the question again and points can be subtracted.

A similar command used in Sony's **BASIC/1** is LOCATE. This command simply allows you to place text where you want it on the screen. "LOCATE 10,2" places the text at Line 10, character space 2.

The proliferation of commands can raise the question of combining some of them. For example, SEEK and PLAY can be combined to become SHOW. However, if you want to play videodisc audio along with IBM video or load a computer image while a video segment is being played, the flexibility of split command functions becomes a positive attribute.

Your IBM's ability to store responses is utilized fully by *Insight*. A student's successful completion of a learning UNIT can demonstrate a grasp of the next two UNITS as well. A student can be rewarded by a series of correct answers with a JUMP command to the next UNIT, which should be more challenging. This is true interactivity, and the people at Whitney Educational Services have spared no ingenuity in providing these tools.

Again, there are numerous commands designed to accommodate videotape players with their long, linear search times. Whole computer programs can be loaded from floppy disk while the VTR tracks down a particular video segment. Even the VTR command holds over in the videodisc chapter to trigger video displays. Once you segregate the tape commands and sort out the videodisc path of the program, its execution is clear sailing.

This type of authoring program falls into a space between the extremely flexible—but arduous to learn—Sony BASIC/1 and attendant video utilities and the Video

Vision Associates' VAI II and *Laser Write* program with its restrictive structure, but extreme ease of use. I cannot go too deeply into each program, because that would take another book, but I am trying to show examples of the opposite ends of the interactive programming spectrum. The Whitney authoring system is definitely aimed at the trainer or teacher who wants true Level 3, external computer control, but does not want to learn a complex language, only a fraction of which is useful for interactive video. One word of advice, however, a good word processor will make this system much easier to use.

APPLE MACINTOSH LASERSTACKS

The burgeoning popularity of the Apple Macintosh microcomputer has naturally caught the interest of interactive software developers. One key component to the Macintosh success has been the HyperCard™, which is currently packaged with each Macintosh Plus, SE and MAC II. This device allows users to create "stacks"—miniprograms—some resembling collections of index cards and others performing special tasks with peripherals (modems, printers or, in our case, videodisc players). Users who create their own specialized stacks swap them, modify them and sell them in "shareware" networks. Software developers have created stacks that have enormous power for a very low investment.

The Voyager Company—Laserstacks

This company has produced a stack that enables the user to use a videodisc as a database. It is an organized catalog of images for students, teachers and industrial trainers. It allows creation of Macintosh icon menus so you can easily mouse-tap your way through existing CAV discs or videodiscs created for a particular requirement.

It offers, for example, "The National Gallery of Art Laserguide," a Macintosh tour of this art collection, using the readily available "National Gallery of Art Videodisc." Works of art can be organized and viewed according to categories, such as artist, nationality, period, style, date, medium and subject. The searchable index includes over 1000 entries, and each entry can be customized by the user with personalized notes. Lists of artworks can be made, saved and printed out. This stack sells for about $50. Two other Laserguides are the "Vincent Van Gogh Laserguide," which sells for about $40 and "Bio Sci Laserguide—A Complete Index," which is priced around $100.

Also offered by The Voyager Company is a toolkit for creating custom interactive videodisc applications. It is called the *Voyager Videostack*.

The *Voyager Videostack* authoring system allows customizing of up to 35 screen icons for videodisc control functions. The user can define a videodisc event—play frame 1000 to frame 2000 using audio track 1—then designate the icons that will activate the event. A feature called "Slide Tray" lets the user access still video frames and then play them back in sequence with descriptions. Automatic installation of any necessary video driver can be loaded into the HyperCard as well. For about $50, you can create a custom interactive video program saved to disk.

Hardware required for these applications is available from The Voyager Company. "Lasercables" allow connection of any Macintosh to the following players with RS-232 ports:

- Pioneer LD-V4200
- Pioneer LD-V6000A, LD-V6010
- Sony LDP-1500, LDP-2000
- Panasonic TQ2024F (OMDR System)

The Voyager Company also offers a large selection of CAV videodiscs which can be used with the accompanying laserstacks, or which can be accessed by a custom laserstack program. Prices of disc sets run from $40.

CONCLUSION

In summary, both integrated and component interactive systems have their advantages. Integrated systems are guaranteed bullet-proof by their manufacturers, but you pay a price in dollars for that security. There are also no standards controlling operating systems or hardware. Few systems allow porting information from one proprietary package to another. Some systems support a variety of authoring languages, while others are locked into an authoring format that restricts the user to match needs with available machine/language functions. The IBM InfoWindow and the Sony View systems seem to be emerging as the most popular choices among software developers. However, major marketing coups scored by companies such as Matrox (the U.S. Army's EIDS program) can quickly establish another "standard" by sheer volume of software demand.

On the other hand, component systems are attractive and cost-efficient. The microcomputer for Level 3 applications and the videodisc player are the two major purchases needed in assembling a system. A user could conceivably upgrade the separately purchased interface "black box" to increase interactivity options. A simple get-started system could be superceded by a more complex system with minimum cost outlay. Of course, the cost saving is accompanied by that spectre of compatibility and the giddy nightmare of making all the components function in concert.

In all cases, the needs of the project should dictate the hardware/software approach chosen. Trying to shoehorn a project into a predetermined system is pure folly.

9 Alternative Disc Technologies

So far, we have concentrated on the standard, garden-variety miracle: the laser read-only videodisc. This is where most of the action is, turning at 1800 rpm, carrying 54,000 frames of storage capacity and available in CAV (constant angular velocity and random access-interactive) with 30 minutes of playing time or CLV (constant linear velocity) with one hour of information per side. The CED disc is a reasonably dead issue because the players are no longer manufactured by RCA, even though discs are still turned out to extract dollars from the 300,000 orphaned CED player owners. A VHD capacitance system created by JVC is not yet making inroads into this country.

JVC's VHD system is growing in popularity in Japan. It now outsells laser videodiscs 65% to 35% in that country. The stylus-record concept that died with RCA's CED discs may come back.

The National Education Corp. (NEC) of Newport Beach, CA, signed an agreement with JVC to provide hardware systems and disc production while NEC creates the hardware for student "Interactive Learning Stations." NEC operates 43 technical schools in the United States and plans to direct its training system toward remedial math and reading programs. Other courses will cover industrial topics such as robotics, electronics, hydraulics and AC/DC fundamentals.

The alternatives lie in future technology. First of all, there is the problem of mastering videodiscs. The average cost of setting up a disc mastering and replication facility is about $27 million. It is understandable, therefore, that most disc mastering is done by a few large companies such as 3M and Pioneer. There are, however, alternatives available for a considerably smaller investment, both for the disc designer and the disc mastering facility.

DRAW TECHNOLOGY

The first we will examine is the DRAW (Direct Read After Write) system. Its main feature is its ability to crank out a videodisc while you wait.

A videodisc designer faces a dilemma, both creative and financial, when his finished master tape goes to a disc mastering facility. A tooling charge of around

$1500 is added to his project for the master disc. If his programming has not been accurate, the mistakes are forever stamped into the laser optical videodisc. Whether he wants one disc, or a thousand, the initial tooling cost of $1500 is still there. If he has made a mistake, he has to have another go, stamping out another $1500 master. This kind of pressure can weigh on the creative spirit. DRAW technology can alleviate this frustration and save money.

Disc creation relies on a basic set of processes. A laser, usually a helium-neon, diode or argon gas laser, burns pits into a rotating medium covered by a laminated protective cover. Beneath that layer lies a heat-sensitive coating that will react to the laser's modulated beam of light. This coating can be tellerium suboxide, polymer-coated rhodium film or polymer-coated aluminum film. Beneath that is a flat substrate made up of ion exchange glass, or polymethyl methylacrylate (PMMA), an acrylic resin. (See Figure 9.1.)

We will look at two standard disc production methods in the appendix at the end of this chapter. The DRAW method of quick disc turnaround strays from the standard process that uses a single laser beam. DRAW uses two beams.

Information in the form of frequency-modulated (FM) signals is stored in a buffer as the disc is being written. Following the writing laser, a second, low-power laser reads the information from the disc just after it is written. The second laser's reading is compared with the information stored in the buffer. If both sets of information match, the recording is a success and the laser moves on to the next track. If there is an error, the track is rewritten.

What this means for the designer is a short turnaround for the disc so it can be checked for programming accuracy. This "check disc" is actually a master. Contrary to disc replication facilities such as 3M and Pioneer where a master disc is stamped and used to create exact duplicates, with the DRAW system every disc is a master. This makes the system awkward and expensive for large runs of duplicate discs, but it makes DRAW ideal for the short run.

Figure 9.1:   The videodisc mastering process.

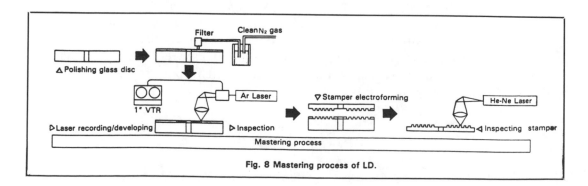

Fig. 8 Mastering process of LD.

Courtesy Pioneer Industrial Video.

ODC Systems

One company built its own market, producing these DRAW systems for disc mastering facilities. In 1982, DiscoVision Associates closed its plant and four founders of the project, John Winslow, Richard Wilkinson, Don Hayes and Ray Dakin, formed their own company, the Optical Disc Corp. (ODC). Between them, they held some 40 patents dating back to the early days of 1969. Their collaboration produced the ODC 610 and 612 videodisc recording systems. (See Figures 9.2 and 9.3.)

This DRAW procedure uses two lasers: a blue argon laser to write the information on the disc and a red diode laser to read the end product. The discs created by the ODC system will play on any videodisc machine. The encoder is designed to insert a number of different disc functions: Level 1 cues, Level 2 cues or film (3–2 pulldown) to video. They have also achieved success with digital data encoding using systems such as LaserData's TRIO, Reference Technologies' DataPlate and compressed audio encoders such as the military's Matrox and 3M's EECoder analog device.

All this precision, such as using a red diode laser for reading data, costs serious money (in the range of $200,000). But if you are establishing a business that will utilize the equipment on a steady basis, the system can be amortized in a relatively short time. If your company requires updated videodiscs, or produces numerous short programs for internal use, the ODC DRAW equipment could fill the bill.

A recordable laser videodisc (RLV), the raw disc offered by ODC, costs only about $50. Any commercial or industrial videodisc player can be hooked up to the ODC 610. A projection of the number of videodisc players operating in the field at the end of the next five years is in the millions. As costs drop and disc compatibility remains, DRAW systems may become even more attractive.

ODC machines have been purchased by a number of duplicating companies and tape editing houses across the country that offer check discs and short runs that cost about $350 per disc.

It is also possible to modify the ODC 610 to work with standard glass substrate discs. A company named Technidisc in Troy, MI, has two of these machines, which require a clean room similar to the Pioneer and 3M operations. DRAW machines can usually operate in any editing suite or office environment.

OMDR System

An alternative to the $200,000 ODC hit is the Panasonic OMDR (optical memory disc recorder) TQ-2026F system, available for about $15,000. You must put up with a few inconveniences for the smaller investment. First, the Panasonic players, recorders and discs are not playable on any other disc system. If you record on Panasonic, you must play back on Panasonic.

The TQ-2026F was introduced in 1984. It allows both record and playback of both motion and still video together with two audio tracks for independent monaural or stereo sound. These audio tracks must be simultaneously recorded and can accept

Figure 9.2:   The ODC 610 videodisc recording system.

Photo courtesy of Optical Disc Corp.

Figure 9.3:   The ODC 612 videodisc recording system.

Photo courtesy of Optical Disc Corp.

"DBX" sound enhancement. The Panasonic outputs are hetrodyne color the same as 3/4-inch videotape and therefore are not designed to be directly broadcast. Standard videodiscs output true color for broadcast use. This feature relegates the Panasonic DRAW system to the industrial realm where both 3/4-inch and 1/2-inch videotape are common coin.

The OMDR machines are designed for record-filing and photo-sensitive archival storage. They can also be used for industrial training, education and medical study. The 8-inch discs can be added on to by recording new images on adjacent tracks. In this way, a record can be gradually filled with frames over a period of time.

The discs spin at the same rate as standard discs, 1800 rpm, and they come in two basic formats: 15,400-frame concentric and 24,000-frame spiral. The 24,000-frame discs can play back 13.5 minutes of real-time motion. (The worst-case frame access time is reported to be .05 seconds, a current world record.) A single, semiconductor laser creates chains of 0.7-micron wide reflective dots in the spiral groove of tellurium suboxide film sandwiched between two layers of polymer film and PMMA substrate. During playback, the laser's power is cut to about 10% of that needed for recording so there is no danger of harming the recorded disc. Panasonic is the exclusive distributor of the discs.

The companion player TQ2027F can interface with a microcomputer through an RS 232 terminal port, permitting Level 3 control. A remote keypad provides Level 2 control. The player lists for about $4000.

LaserFilm

Another proprietary disc technology comes from the McDonnell Douglas Electronics Company. It is called LaserFilm and records up to 30,000 frames with compressed audio capability and comes with a fully configured interactive disc player and videotape formatting system. By utilizing photographic film as its disc medium, the user can master discs and duplicate the result at a low cost. The Video Listing Service (VLS) real estate system has committed to the purchase of 20,000 of the players that will be coupled to IBM PCs for showing properties to prospective buyers. Updated discs can be created on a weekly basis with the inexpensive process.

McDonnell Douglas delivered an Operation and Maintenance (OMIS) System to NASA in 1987. The system makes use of 1-inch video monitors designed to be worn by space station maintenance personnel. The monitors are linked to a computer via a voice-activated link and the computer controls the LaserFilm videodisc library. Repair instructions, tool requirements and system information can be projected on the faceplate of the repair person's helmet, on demand.

The LaserFilm players are manufactured in Japan by Sansui.

Market analysts project that LaserFilm discs could command 8.8% of the Japanese optical data storage system market by 1990. The McDonnell Douglas system is making inroads in the U.S. market with companies that prefer to keep their videodisc production in-house, so frequent updates can be made.

LV-200 System

Another challenge to the proprietary videodisc market has been made by the TEAC Corporation. It has introduced the LV-200A videodisc recorder and its companion player.

This system can record and play back one hour of video or 108,000 still pictures on a 12-inch disc, using the "Standard Play Disc" (CAV equivalent), or two hours of video on the "Long Play Disc" (CLV equivalent). (See Figures 9.4 and 9.5.)

The recording features are extensive, such as direct loading of 35mm slides onto the disc, using the VY-420 Film Video Processor (see Figure 9.6), for archival storage, interval recording of events with preset times ranging from one second to 99,999 seconds between exposures directly onto the disc, a built-in real-time clock for inserting second, hour, day and month over recorded video frames, and programmable preset automatic recording using the LV-200A's 8K memory with a battery backup.

With a full-function keypad, Level 2 programming is possible up to 150 steps. This function also takes advantage of the battery backup which is good for up to 10 years and assures that a program will not be erased on power-down. An RS-232 computer port control allows control of either the recorder or player.

TEAC has also created a utility language for computer-controlled image search and start-stop video points. This program (called *MUMPS*) runs under standard MS-DOS which allows any IBM or compatible to control a TEAC Laser Videodisc when

Figure 9.4:   The LV-200A videodisc recorder.

Photo courtesy of TEAC Corp.

Figure 9.5:   The MA-200W videodisc player.

Photo courtesy of TEAC Corp.

Figure 9.6:   The VY-420 Film Video Processor.

Photo courtesy of TEAC Corp.

coupled to the appropriate TEAC interface circuit board (VA-414, VA-415 or VA-416, depending on the type of computer port used).

One of the most recent uses of the TEAC system in the public sector was by the Dayton Hudson department store in Minneapolis, MN. A 32-screen videowall created by Electrosonic (based in Minneapolis) makes use of the TEAC technology. Fashion shows, shopper information and product marketing are some of its applications. The need for frequent updating from videotape required a disc system that could be recorded in-house at low cost. This enabled short-term, sales-oriented programs to be assembled in a few hours for feeding into the 32-screen, computer-controlled matrix.

RECORDABLE LASER VIDEODISCS

Recordable laser videodiscs (RLV) are showing up in a multitude of forms, mostly related to black-and-white and color document filing and storage. While the United States favors databases, Japan, for instance, is very facsimile-oriented. This is partly due to the 3000 characters in their written language, which requires most documents be hand-written.

Optical discs can save considerable storage space over conventional film libraries. The systems producing these discs in Japanese companies are typified by the Toshiba DF-2100R.

This compilation of separate machines includes a keyboard for system control, a scanner that reproduces any document and feeds the information to a RLV optical disc recorder and a copying device that can retrieve any document from the disc and produce a thermal paper copy. Each 12-inch optical disc can store up to 10,000 single-page documents. A 14-inch green CRT provides visual verification of all documents to be scanned and/or copied.

This RLV disc production method is achieving the original goal of videodisc designers—creation of cheap, readily accessible discs for many different uses. Hopefully, the standardization of disc formats will continue so this infant market can develop new hardware without continually reinventing the wheel. With RLV discs, interactive video can be even more widely used in industry and education.

ERASABLE OPTICAL DISCS

Another alternative disc technology is a true breakthrough that will affect the future of media storage. For years, magnetic media storage has been confined to variations in tape, floppy disks and hard "Winchester" disk systems. As more storage capacity has been required, manufacturers have pressed these established formats to their limits. Tape reels are the standby of mainframes. High-speed, streaming tape is not practical for the microcomputer, because the tape storage devices cost more than the micros.

Microcomputer programmers began their storage of data with what they could stuff in volatile random access memory (RAM), but the data vanished when the power was shut down. Tape cassettes in ordinary tape recorders followed, but were terribly slow. Floppies came next, but the 8-inch size proved unwieldy. Mini-floppies at 5 1/4-inch were more portable, and data could be packed up to 1 million kilobytes per side. Three-inch disks followed that offered the same capacity.

Aluminum disks coated with magnetic medium, spinning in airtight chambers, were the next major step. These "hard" disks are now required for many of the memory-gulping programs offered in today's applications software. Their capacities range from 10 on up to 100 or more million bytes of data. But they are also volatile and must be backed up onto tape or floppy disks for safety reasons (e.g., in case of a disk "crash" due to a power fluctuation or accidental bump while data is being written).

The videodisc has long been touted as an excellent storage medium for images and computer data, but its greatest drawback has been its inability to change any of the data recorded in its billions of micron-sized bits. Once laid down, the information was there to read, but not to alter. That has changed.

New discs make use of magneto-optic technology. The disc is coated with a layer of magnetic media and in the write mode, a small DC magnetic field is applied to the media. Light from a laser is focused on a 1-micron spot to rapidly heat the spot of media. This process reverses the magnetic polarity of the spot and a bit of data is recorded. A lower-power laser can then read the bits after they are laid down.

Want to erase the data? Run the laser over the same track-full of data and *reverse* the magnetic polarity of the bits back to their original state, and then the track can be rewritten. The amount of data written depends on the area covered by the laser diode. A 1-micron-sized bit burned into the medium will allow a density of 5×10^8 bit/cm^2, or roughly 300 to 500 million bytes of data. This is equivalent to 50 to 60 hard disks, or some 700 floppy disks, on a single 5 1/4-inch erasable optical disc. Since the data is being written with a laser, the need for either contact with the disc surface, as with floppy disk drives, or even for low-flying magnetic heads in sealed hard disk drives, is eliminated.

Besides 3M, both Sony and Verbatim have announced new magneto-optic erasable discs. Sony bought rights to an erasable optical memory disc from Energy Conversion Devices, Inc. of Troy, MI. The disc allows the storage of 170 million words. Verbatim, a subsidiary of Kodak, demonstrated a thermo-magneto-optical disc in 1985. The Verbatim disc is 3 1/2 inches in diameter, capable of storing 40 to 100 million bytes. The drive should sell at about $300 and the discs at $30 to $50 each.

As of this writing, the industry has swung sluggishly toward this new storage technology. Naturally, with the increased density of data, the mechanism used to read and write to the disc must be particularly precise. Also, the writing of data can tolerate the least amount of surface contamination, suggesting a sealed system not unlike the hard magnetic disks currently in vogue. The mechanics required to make use of the erasable laser disc present formidable challenges.

Erasable disc technologies are developing along with research into mass-produced chip memories. A chip memory is passive, residing on a circuit board, and yet is not volatile in the manner of RAM. No mechanics are required. No special environment is needed in this era of shrinking electronic components. The optical disc wins hands down in terms of sheer capacity, however, and considering today's memory-hungry software and huge database management systems, the erasable optical disc seems to be a sure winner.

CD-ROM

At this writing, laser software for computers is showing up in the marketplace in the form of the CD-ROM (compact disc, read only memory). These are vast data banks that can be read by a computer connected to a CD-ROM disc drive system. The first offering was Grolier's Electronic Publishing's Academic American Encyclopedia—21 volumes on a disc for $199. The data was accessible through software designed by Activenture of Monterey, CA, a company headed by Gary Kildall, creator of the CP/M computer language and founder of Digital Research.

WORM SYSTEM

A hybrid of the erasable disc and CD-ROM is the WORM (write once read many) system. The disc can be written to once and then read, but not erased.

Optimem in Sunnyvale, CA, has produced a 12-inch WORM disc and player capable of storing one gigabyte, or 400,000 typed pages.

CD-I, CD-V AND DVI

Finally, we come to a series of new technologies that are developing as this is being written. Some cynics refer to these technologies as "Vaporware"—or highly touted experiments that do not have actual applications. Only the marketplace and the test of time will decide their survival. For the most part, they form a new alphabet soup to be digested.

First, there was CD-ROM, which exists in a number of forms and is described earlier in this chapter. Next, came CD-I (compact disc, interactive). This system is being developed as a consumer product tied to a "talking book" type of hardware. The concept was developed by Phillips in the Netherlands and makes possible interactive novels, games and educational programs that can be played on machines which can have their own screens or can be attached to TV sets. These compact discs will not play on any present CD players. Text, still photos, graphics and reduced "window" full-motion video are possible. A number of manufacturers are running "start-up" programs to see if there is a market.

Another venture is the CD-V (compact disc, video). This consumer software uses the music capabilities of the popular compact disc, then adds five minutes of full-motion video to the 20 minutes of music without video. Actual CD-Vs are available, which take advantage of the music video market sparked by MTV. Pioneer, Magnavox and Yamaha have produced "Combi" players that play 12-inch videodiscs, CDs and CD-Vs—all costing around $800. So far, lack of software has inhibited the CD-V's growth and the production of straight CD-V players of the quality of the current CD playback systems.

Another entry in the alphabet sweepstakes is DVI (digital video interactive) created by the General Electric/RCA Laboratories in Princeton, NJ. This system allows production of up to one hour of full-motion video from a standard CD-ROM disc.

The key problem here has always been one of storage capacity. A standard screen of analog video requires roughly 600 kilobytes of digital data per frame. At that rate, a CD-ROM disc could only store about 30 seconds of digital video. And, since a CD-ROM disc only reads data at about 150 kilobytes a second, any video motion could not be seen in real time.

Using compression algorithms and special chips, DVI crams all the data into the given space on a CD-ROM disc. A second set of *de*compression algorithms reads the data back out again. What all this means is that a microcomputer could output text, graphics, sound and full-motion video from its DVI software/hardware package. At present, no applications are on the market, but anticipation is running high.

CONCLUSION

Alternative disc technologies are expanding the horizons of this medium. DRAW discs allow inexpensive production of optical discs. Recordable laser videodiscs such as the TEAC and McDonnell Douglas systems make possible in-house disc production. The storage of data is being revolutionized through erasable discs and the CD-ROM. Finally, CD-I, CD-V and DVI technologies are stretching the capacities of our compact disc media. Through seeking out alternatives, we constantly revise and explore our concept of the disc medium.

APPENDIX: VIDEODISC MANUFACTURING

The laser videodisc is composed of billions of pits, each pit being 1-micron long, 0.4-micron wide, 0.1-micron deep and spaced 1.67 microns apart, center to center. The laser beam's focus point is 1.5 microns in diameter as it sweeps down the track full of bits, translating the fluctuations in the bits' reflectance into modulated signals that are, in turn, translated into data or pictures. The beam is allowed a 0.5-micron "slop" to compensate for tracking error, which is compensated for with an alternating electrical current. (See Figure 9A.1.)

A micron is 1-millionth of a meter, or 0.000039 of an inch. A human hair is about 10 microns thick. This is beyond tiny and requires incredibly precise manufacturing techniques.

All manufacturing must be conducted in a "clean-room" atmosphere. This means all air is filtered and all clothing worn in the area must be covered by special white "bunny" suits that are designed to be lint-free. Shoes are encased in plastic boots and any beards or facial hair must also be covered. Rubber gloves are worn at all times. The human body is considered chemically "filthy," even after a dozen showers. We are constantly shedding skin bits and hair follicles. Bacteria are constantly present on our skin surface. A particle the size of an ant's eyelash can cause a signal drop-out on a videodisc. The clean-room environment is absolutely necessary for final disc mastering.

"Check discs" are offered by most mastering houses. These discs are less expensive because they are created either with DRAW technology or a recordable laser videodisc that does not require the clean room. Audio and video quality on check discs are no match for the precise system we will look at here, but are quite good for seeing if the programming was correct and if the disc performs as it should.

Once the tape has been checked, edited and encoded with required disc data, the tape information is transferred to a glass master disc. This disc, about 1/4-inch thick, has been precisely polished to ensure a perfectly flat plane. It is then washed both chemically and supersonically. A thin coating of photoresist is applied. This coating is applied under photo darkroom-type light quality and sensitizes the surface of the glass disc to intense laser light. The disc is then taken to a real-time disc cutter.

The composite video and audio signals are fed into an optical modulator that switches the high-powered laser beam on and off—on creates a pit, off allows the spaces between pits. The beam is projected onto the sensitized surface of the disc, 55 millimeters off the center, and as the disc revolves, pits are burned into the photoresist. The pits are recorded in a spiral moving outward from the center with a distance between tracks of 1.67 microns. When the cutting is complete, the photoresist is developed much like a piece of photographic film.

Next, the disc is placed in an evaporation chamber and the air is pumped out of the chamber, creating a vacuum. A fine layer of nickel is applied. Additional chemical processing transforms the nickel surface into a complete metallic pattern of the pits. These pits are so close together that they play beneath the laser beam at a rate of up to 10 million per second.

One more process is needed for the glass disc. It is dipped into a vat of nickel that is electroplated to the surface of the nickeled photoresist. The actual master

Figure 9A.1:   Videodisc construction.

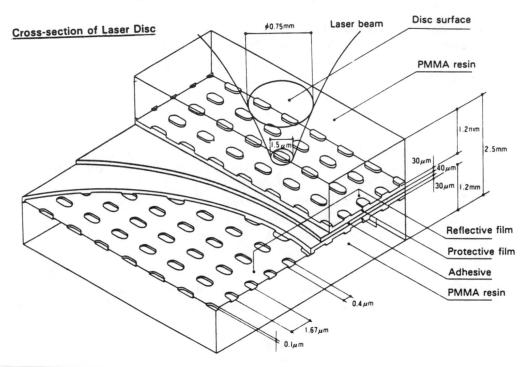

Cross-section of Laser Disc

Laser beam · Disc surface · PMMA resin · Reflective film · Protective film · Adhesive · PMMA resin

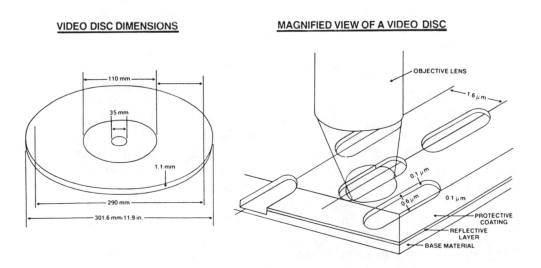

VIDEO DISC DIMENSIONS

MAGNIFIED VIEW OF A VIDEO DISC

Courtesy of Pioneer.

disc, or "stamper," is created. A microscope is used to align the center of the disc in relation to the outer track. This correct alignment allows the large center hole to be accurately cut. The stamper is now ready to replicate discs.

The replicating line is the next destination for the stamper where it is installed on an injection molding machine. The clean-room environment is still maintained. Transparent, acrylic resin is squirted into the disc form on the machine and the stamper is lowered to make an impression. The resin is cooled in a few seconds and a clear replication is made. This operation is repeated for as many discs as are ordered.

The clear discs are transported along an automated conveyor line and cleaned by ionized air jets and passed through an identification stamper that presses a unique identification number into each disc. A robot arm then picks up each disc and stacks it vertically in a special "caddy" and the discs are placed on a carousel. The carousel transports the discs into another, larger, evaporation chamber where a layer of aluminum is vapor-deposited on each disc.

Actually, a videodisc is made up of two discs bonded together, information side-to-information side. Each aluminized disc represents half of a finished disc. Once out of the evaporation chamber, the discs are coated with a protective resin. The dynamic weight balances of side one and side two are measured and trimmed. They are bonded together to form the finished disc. At this time, the discs leave the clean-room environment for the first time for labeling, extensive testing and quality control. Discs are randomly selected and tested for frame access, programming, video and audio quality.

The stamper has great longevity. Pioneer has had runs of 20,000 and the stamper was still in good shape.

The above information relates to Pioneer's process. 3M does it a bit differently. They take a plastic substrate—a ready-made disc—with a blue, light-sensitive coating on it. The stamper touches the surface and the coating is exposed to ultraviolet light. The disc is rinsed in developer and then aluminized. After that, the processes are almost identical, except that the resin coating is somewhat different.

Pioneer uses a "roller coater" which applies a protective liquid, while 3M uses a plastic film that seals the surface. In essence, Pioneer uses injection molding and 3M uses the photo-polymer (PP) process. Sony used the injection molding process in Japan, but now has 3M handle all its replication in the United States.

10 Conclusions

We have come a long way. The computers and disc players are back in their cartons awaiting return to the nice people who loaned them to me. The stack of manuals is only a bleary-eyed memory. Now is the time to assess what I've learned about the DISConnection—this technology that is only about 10 years old.

People involved in some facet of the connection say, "It's exciting!" And yet, trying to pin down that excitement is like trying to nail Jello to a tree. The "exciting" statement is almost always followed by ". . .and it's damned frustrating." There is a "Catch-22" reality to the search for interactive video.

EDUCATIONAL AND TRAINING APPLICATIONS

Bruce Gjertsen is a consultant, working for a microcomputer consortium called Micro Ideas. This group is concerned with the use of microcomputers in schools and the development of educational software in the state of Illinois. More than 100 schools belong to Micro Ideas and educators take advantage of the group's seminars, workshops, software development and technological know-how. Bruce and his cohorts have also entered the realm of interactive video. What more natural application than in the formative classroom?

Bruce sees videodiscs as the exciting new media, and *business* is leading the way. He claims that's where the money is. "Education has no money and is very conservative when it comes to new technology. In our public and private schools, we are talking primarily Apple computers. These machines have a limited amount of RAM memory and this limits the power of any authoring system. There is a huge installed base of Apples in the schools, so that is our basic interactive tool. As you know, most of the interfaces and authoring systems being designed today are compatible with IBM systems and the more powerful Macintosh computers."

In effect, Bruce is trying to promote a way of teaching that is dying off as he makes converts.

He also claims that there are some new companies out there trying to meet educational needs. "This kind of flexibility is essential for the education market. Teachers' learning curves are very time-intensive. They can't spend a lot of time try-

145

ing to learn new command sets in order to create CAI, or interactive video lessons. It's almost impossible to get a school district to roll over its Apples for IBM PCs or Macs unless a real benefit can be shown at the bottom line. With new equipment being constantly introduced, people have become stunned. A technological paralysis has set in."

Business, however, is taking to interactive video for training in a big and expensive way.

As we've seen in the previous chapters, many large corporations such as Ford, General Electric, AT&T, IBM and General Motors have made large investments in interactive video. Major training programs exist in these and hundreds of other companies that utilize videodisc systems, and the numbers are growing. This growth has prompted introduction of courseware written for the business community and alternative technologies such as the systems offered by TEAC and McDonnell Douglas. These advances help bring interactive video into the companies where the cost of development was prohibitive and inflexible.

The military continues to be a major driving force in the dissemination of videodisc training programs. The high-tech nature of modern military equipment means that the Armed Services must depend on the most effective training methods possible for its high-turnover work-force. Everything from tank gunnery to missile rocketry is being taught through interactive videodisc training. This reliance on interactivity must make the same demands on military subcontractors that prepare training programs to accompany their systems. As these programs become more common, their trickle-down effect will permeate the rest of our industrial complex.

Interactive Video as a Training Success

The case histories I have set down all seem to point to the fact that interactive videodisc training fosters a greater interest among students, promotes better concentration, produces higher comprehension test scores and reduces subject learning time. Whether the student is in a public school, wielding a wrench in a Ford classroom or in a fighter cockpit simulator lining up his sights for a Sidewinder missile launch, interactive videodisc technology is improving the way we learn and how well we retain what we are taught.

If training in the business community is becoming more effective through interactive video, the use of videodiscs in the marketplace is creating some bottom-line results that can't be ignored.

POS SYSTEMS

Point-of-sale (POS) video kiosks are popping up all across the country and are changing the way consumers shop. They are also changing the way retailers and service vendors view their marketing mix.

Bob Sandidge, president of New Orient Media in West Dundee, IL, and a designer of POS systems, did his homework before he entered this new technology.

According to Mr. Sandidge, in 1984, there were approximately 1000 POS systems in place. Now, there are forty times that number.

This is the beginning of a major trend. Now, a manufacturer can have a 2-foot square piece of real estate in a shopping mall that is a *real store*. Not only do POS kiosks conserve floor space, they also reduce the number of personnel usually necessary in regular stores.

Kiosks are very sophisticated systems. A shopper can view products, make selections and even place orders with a credit card. Some kiosks even offer a hardwired modem that allows shoppers to place orders over a phone line. Modems also provide the retailer with a convenient way to update the computer software in the kiosk from any phone in the world.

The POS kiosk is tireless, does not need a coffee break and effectively places more product in front of more people all the time.

A study conducted by ByVideo in Sunnyvale, CA, has shown that where competition for shoppers' attention is considerable, as in a shopping mall, these bombarded souls are willing to spend from three to six minutes at an interactive videodisc kiosk electronically thumbing through a message. If the information is interesting to the shopper, browsing time increases to 11 minutes.

The appeal of these systems, particularly the touch screen models, is evident with both large and small businesses. Following are just a few examples. Sears recently opened numerous kiosk networks. Cuisinart, Inc. has placed more than 200 single-product–oriented kiosks in department stores since late 1983. Realty systems such as the Comsell Realty Showcase allow home buyers to study houses via videodisc after feeding requirements into Comsell's computer database. A small liquor chain on the East Coast has placed kiosks in its stores to help customers select wines by typing in dinner menus.

The applications are numerous and cover a wide economic span. Bob Sandidge has an interesting vision concerning the future of point-of-sale systems:

"I can easily imagine this arcade of POS touch screens, each one vying for the shopper's attention. The screens will offer shopping, entertainment, tickets and reservations and service information."

If that vision seems a bit chilling, remember the last time you had a confrontation with a surly discount store employee or a snotty reservation agent. For no-fuss transactions where there is little compulsion to kick the tires or sniff the leather, then the interactive video store comes well into its own.

HARDWARE AND SOFTWARE

The interactive video hardware and software market of today is not unlike the personal computer market was from 1978 to 1982. Every week, a new machine appeared. Everyone wanted a piece of the action and felt their particular system was *the* answer. The personal computer burial ground is littered with the bleached bones of undercapitalized companies; companies caught between product upgrade and swollen inventories; companies that lost suppliers, shelf space, dealers and their shirts. The lesson was not lost on future high-tech innovators.

The "technical paralysis" noted by Bruce Gjertsen earlier has been created by the current glut of "interface systems." Pick any educational, audiovisual or video trade publication and turn to the "New Products" section. Chances are you will find a new company offering another videodisc interface and authoring system. Slick, four-color ads, mailing sheets and product reviews extol the latest creations. Potential users take up their pens and fire up their word processors to ask for more information and place orders for these wonders.

Unfortunately, many of these companies are small in both manufacturing capacity and number of personnel. The computer lesson has made "caution" the watchword among these entrepreneurs. High-tech venture capital has dried up. In many cases, the company behind the four-color ad blitz is really a mom-and-pop shop. Consequently, when one of these companies goes under, there's barely a ripple in the pool.

This does not help the company or school district that has finally decided to plunge into the "Great Interactive Experiment." Orders are placed after much agonizing. There is no real way to test interactive interfaces and software put out by these dozens of companies. No magazines review these systems as they do in the personal computer realm. No stores stock these devices. Hands-on evaluation is the only judging ground.

STANDARDIZATION

As the field of interactive video grows, standards will eventually surface. Computers went through the same problem. First, there was the CP/M operating system and then came Apple DOS 3.3 and MS-DOS. Today, OS/2 is slated to be the industry standard for serious business use, while UNIX and Apple Macintosh are serious alternatives. PILOT, PILOT Plus and SuperPILOT were recognized as semi-standards. Many systems were based on the PILOT series of language upgrades. Today, IBM InfoWindow and Sony View systems are attracting programmers and courseware writers.

The "consumers" of interactive videodiscs are not primarily individuals, as it was with personal computers. The dollars for this industry are being coughed up by committees, corporations, organizations and a considerably literate portion of our population. A shakeout will occur, driven by these tough purchasers. This kind of support can force a standard as it did in the computer market.

The majors, however, are fighting their own battle. Market share is still the name of the game and videotape cassettes still account for the largest share of consumer video purchases. Volume means dollars to pursue the interactive market and create these standards. Pioneer chose to build the CLD-1030, a disc machine that plays both videodiscs, compact audiodiscs and the new CD-V entertainment discs. Other player manufacturers are following suit—introducing the superior playback of videodiscs to consumers on the coattails of the popular CD.

NEW TECHNOLOGY

The laser-read disc is also seeing a new market as the CD-ROM. This compact disc, read only memory, medium is being manufactured as an answer to data storage on a huge scale. The CD-ROM disc can store upwards of 20,000 magazine pages, including photos, on a single side. The disc cannot be written to after it has been stamped, but many storage applications do not require constant updating, just simple storage. This technology is on the immediate horizon.

The WORM (Write Once Read Many) disc is another potential money maker. The industry is trying to catch up with methods to utilize all this new technology.

THE FUTURE

The training and point-of-sale uses of videodiscs have helped keep the medium alive and have advanced its technology in fits and starts by a few large and dozens of small companies. Its advantages are apparent, but so are its drawbacks: cost to produce, low sales volume and lack of standard systems. As with most high-tech innovations, the product seeks its level in the marketplace. Bob Sandidge had an illustration for this point: "A university in Mexico built a new campus. They created a series of buildings to house classes and faculty. Eventually, everything was complete—except for the sidewalks. The developers waited for a few months and eventually the paths taken by students and faculty from building to building became evident. Then the developers laid sidewalks to conform to those paths."

Perhaps the videodisc developers are watching the paths as you and I use what charts we have to learn how to interact with videodiscs.

Glossary

This glossary has been compiled over a period of 10 years as I have worked with both computers and video productions. In order to help learn my craft, I collected words that I didn't understand and also words I had to look up in other glossaries.

A glossary is a living document that should provide more than a few hints to the jargon in a text. A good glossary is necessary for a book such as this. I hope this compilation of words will help the reader deal with video and computers even beyond the scope of this book.

A and B editing: The preparation of material recorded on videotape, using A and B rolls of tape carrying the same material on each roll and editing back and forth between the rolls, inserting the scenes sequentially in building block fashion. A and B editing is most valuable when a number of dissolves and special effects are anticipated. "B" rolls are then recorded from the master footage for intercutting with that footage. The loss of one generation of quality from the A to the B roll is generally insignificant so as not to be noticed in the final print.

above the line: Budget category that includes the artistic or creative elements—primarily non-technical personnel and activities.

access: Going to or reaching an item on a videodisc, or videotape.

arc of good location: The portion of the geosynchronous orbit (22,300 miles above the equator) that provides optimum coverage of the United States.

audio track: That portion of the videotape on which audio information is recorded.

automatic gain control (AGC): An electronic circuit that adjusts the incoming signal to a predetermined level; an automatic volume control.

automatic level control (ALC): Electronic circuit that adjusts the incoming video signal to a predetermined level. Same as AGC with audio.

automatic light control: Vidicon camera control that automatically adjusts the target voltage to compensate for variations in light levels (auto-exposure control).

available light: Source of lighting in the scene to be taped, usually referring to natural lighting and practical sources that normally exist to illuminate the area.

baby: 750-watt spotlight.

baby legs: Low camera tripod.

backing: The plastic ribbon, usually mylar, of both audio- and videotape, which is oxide coated.

back light: Light placed behind objects in the scene and pointed toward the camera to provide a rim of light on the object, which separates it from the background and gives the scene a feeling of depth.

backporch: Five-microsecond portion of composite video signal lying between the trailing edge of horizontal sync pulse and the trailing edge of corresponding blanking pulse.

backspacing: Reverse cueing technique.

backtiming: Reverse cueing technique for editing backspace used in electronic editing.

bandwidth: Number of frequencies contained in a designated channel.

barn doors: Metal flaps that attach to the sides of a lamp housing to control the light being cast into an area.

barrel distortion: Distortion of a scene by a wide-angle lens; everything near the edge of the scene appears rounded and out of proportion when objects are close to the camera.

beam: A semi-coherent flow of electrons.

beam splitting: Method of dividing the color components of the image so they can be cast upon more than one vidicon target area (or tube); used in two-, three- and four-tube color cameras.

below the line: Technical and production costs as indicated in the program budget—includes production equipment and technical personnel.

bidirectional: A directional microphone that accepts sounds from only two directions and attenuates sound waves from any other direction.

bird: Slang for a communications satellite.

black clipping: A video control circuit found in cameras and VCRs that regulates and contains the black level of the video signal so it does not disturb or appear in the sync portion of the signal.

black level: Minimal television voltage signal establishing blackness of the transmitted image. The bottom level of the picture signal, below which the sync blanking and other control signals do not appear as picture information.

blanking: Suppression—the process and period of time the process takes during the scanning of the raster area when the beam is shut off. Line blanking takes place when the beam is returning from the end of one line to begin another. Also the period of time in which the beam finishes scanning one field and retraces its path to the top of the raster to begin scanning the next field.

blanking signal: The pulses added to the video signal to indicate that the signal from the beam to the target area should be cut off.

block: To work out talent and camera positions within a scene before taping.

bloom: Undesirable video picture caused by excessive light saturation.

boost: To turn up or make the video/audio signal stronger.

branch: The action of routing a user to a particular part of a computer program or video-disc segment based on the user's responses.

brightness ratio: An indication, expressed as a ratio, of the difference between the whitest and blackest object in a scene. Too wide a brightness ratio can lead to an unacceptable video picture.

brightness value: Luminance—the relative brightness of a particular object in a scene. The point on a gray scale at which the object is between absolute black and absolute white.

broadband: Ability of a circuit to operate over a relatively wide range of frequencies. Cable **TV** is often referred to as broadband communication.

burn (burn in): Overbright images that are retained on the surface of the video camera tube (can be removed by aiming the camera at a brightly lit card).

bus (audio/video): One computer channel of an audio or video mixing system, including inputs, gain controls, and an output. Two or more buses are required for video signal switching.

bus (computer): A path of signals along which data travel to and from destinations and sources. Every computer has at least three buses: address bus, data bus and control bus.

camera: The eye of the video system capable of absorbing the light values of a scene and converting them into a corresponding series of electrical impulses. Using a cathode ray tube, the light values are changed into voltage variations that are used to recreate those values on another cathode ray tube in the video monitor.

camera chain: The camera and its associated electronics (camera, control unit, power supply).

cannon connector: A brand of audio jack that features three heads: two for the signal and one for overall system grounding. A very secure type of connecting jack.

cannon plug: Special three-prong plug and connector jack that locks male and female plugs together with a small latching device. Used on all professional video equipment.

cans: Earphones.

capacitor: A component used in audio and video circuits to store and release voltages within the circuit.

capstan: A rotating shaft on the VTR that is turned by a motor and which, in turn, governs the speed of the tape from reel to reel.

carrier frequency: The particular wavelength of a certain frequency on which a signal is impressed for transmission in a coherent fashion to a receiver. Here it is stripped of its carrier frequency, amplified and reproduced (either audio or video).

cathode ray tube (CRT): A vacuum tube with a cathode and heater element at one end capable of producing electron beams. The beams flow down the length of the tube where they either hit a phosphor coating on the face of the tube and make it glow or hit an oxide coating and produce a voltage. Both a vidicon and a video monitor are cathode ray tubes. Phosphor coating is for monitors, voltage is for vidicon camera tubes.

C-band: The range of frequency, from 4 to 6 Gigahertz, on which most communication satellites receive and transmit signals.

character generator: A device that electronically displays letters or numerals on a video screen.

charge coupled device (CCD): A light sensitive chip that replaces video tubes in cameras.

cherry picker: Motorized high-angle camera crane position with an operator bucket.

chip chart: Standard black and white scale test chart for video camera alignment.

chroma keying: 1) The electronic introduction of a color background into a scene. Unlike black-and-white keying, color is present and the operator can adjust the color at the keying unit. 2) The keying in of an object against an established background—two images fused together electronically. A blue-green background against which an object is placed for keying against another scene is called chroma key blue.

chromaticity: A subjective evaluation of the hue and color saturation of an object.

chrominance: Chroma—the hue and saturation of an object as differentiated from the brightness (luminance) of that object.

chrominance signal: That portion of the total video signal that contains color information. Without the chrominance signal, the picture would be black and white.

circuit boards: A plastic board that holds the electronic circuits necessary for the operation of a computer's hardware. In common use, circuit boards are also called "cards" and are inserted in a computer's expansion slots in order to add operation capabilities to that computer. The "mother board" is the main computer circuit board.

clamper: An electronic circuit that sets the video level of a picture (or audio) signal before the scanning of each line begins to ensure that no spurious electronic noise is introduced into the picture signal from the electronics of the video equipment.

clean edit: An electronic edit of a video picture that has no noise, distortion or other disruption when the signal changes from picture 1 to picture 2.

clip: To cut off sharply.

clipping: A circuit that removes the positive and negative overmodulations of a composite video signal so they will not interfere with the picture sync information.

closed circuit: Distribution system using wires or microwaves to connect receiving sets to transmission equipment (CCTV).

C-mount: Standard screw thread size for adapting 16mm lenses to video cameras. Most C-mount lenses are single focal length (9mm, 25mm, 150mm, etc.) rather than zoom lenses that come with the video camera. C-mount lenses are used for long telephoto or extra-wide angle shots.

coax cable: A one-ground, one-conductor cable that can carry a wide range of frequencies as far as 1000 feet with little or no frequency loss.

coaxial connector: A specially designed cable connector used in cable TV and other 75 OHM cable applications.

color background generator: An electronic circuit used in chroma keying to produce a solid color background of any desired hue or saturation.

color bars: SMPTE standard test bars; electronically generated bar-shaped videotape leader color pattern to match playback to original recording levels and phasing. Usually accompanied by a 1000Hz audio reference tone.

color dissector tube: Color tube; chrominance tube. A cathode ray tube designed to separate a scene's hue and saturation values into its red-green-blue (RGB) components for electronic encoding as part of the color video signal.

colorizer: Electronic circuit used to generate a chrominance signal in relation to the gray values in a black-and-white signal. Each gradation of gray from black to white is assigned a color value. The result is an artificially colored picture that does not truly represent the color of the scene.

color killer circuit: An electronic circuit used in a VCR to suppress the 3.8MHz color carrier frequency when a black-and-white tape is being shown. Without the circuit, the color signal will cause picture noise in the black-and-white image.

color phase: The proper timing relationship within a color signal. Color is considered to be in phase when the hue is reproduced correctly on the screen.

colorplexer: Encoder—electronic circuitry that processes three separate color signals (red, green, blue) coming from the pickup tubes into one composite encoded color video signal.

color subcarrier: The carrier wave on which color signal information is impressed. Contains burst signal and alternating phase color information (3.58MHz).

color sync: A control signal necessary for the operation of two cameras, segs and monitors; consists of a 3.58MHz burst (which sets the color phase and placement before each line is scanned) and a 3.58MHz color subcarrier.

co-microprocessor: A microprocessor that supplies functions not supplied by the computer's primary microprocessor. Example: an Intel 8087 processor provides math functions beyond the capability of the IBM PC's 8088 processor. A co-microprocessor can supply functions such as graphics, math, music and voice recognition.

compatible color: System that produces a color signal that can be reproduced on either a color or black-and-white set. Luminance (black-and-white-values) and chrominance (color values) are separated so luminance does not rely on chrominance to reproduce a picture signal on a monitor or TV.

component: Any portion of a total electronic system.

composite master: An original program produced by editing various portions of other recordings onto a new reel of tape. In electronic editing, the resulting tape is one generation away from the master materials from which it was recorded.

composite sync: The total sync system containing both vertical and horizontal scan controls.

composite video signal: Video signal containing both picture and sync information.

compression: Audio term similar to video clipping is the automatic adjustment of volume variations to produce a nearly consistent level of sound. Elimination of audio overmodulations produces a sound track lacking in dynamics—it is never soft or loud, but always at the same level.

contrast range: The range of gray between the darkest and the lightest value in any scene. Expressed as a ratio of light to dark, such as 20:1, and used to evaluate a picture on a monitor screen.

control track: The lower portion along the length of a videotape on which sync control information is placed and used to control the recording or playing back of the video signal on a VCR. Editing a tape that has only control track instead of time code is very difficult. Master tapes (edited) have control track information instead of time code. Time code must be added to master tapes in order to have editing flexibility in combining master elements.

cookie: A cutout screen placed before a light source to cast random wall shadows. Light source should be a focused source through a lens (an ellipsoidal lamp is ideal, or a slide projector can be used).

corner insert: A second video picture signal inserted into the first video picture signal. Corner inserts are achieved by halting the horizontal and vertical scanning of the first picture in a predetermined area and inserting the second picture scanning portions into that area.

crash edit: A brute force, electronic assemble edit that may leave a slight glitch or distortion at the edit point on playback. See *glitch*.

crop: Camera framing; or framing of an object to exclude some picture information.

crossfade: To fade out one video signal and fade in another as a simultaneous movement. See also *dissolve*.

cross talk: A spillover of sound from a line to an adjacent line or on an audio/videotape from one layer to an adjacent layer of tape (also called *bleed-through*).

cueing: Presetting a record, transcription or a tape on the first playback machine for immediate starting.

cue track: An area that runs longitudinally along the videotape and carries audio and editing information in the form of reference pulses or time codes.

cut: To instantly replace one scene with another.

cut-away: Videotape shot of an interviewer that may be interspersed during the editing process to avoid a jump-cut editing of the interviewee. Any shots (close-ups, reactions, etc.) that can be used to break up long scenes or allow transitions to other parts of the program.

cutoff (TV-cutoff): Section of the transmitted image that is hidden from viewers by the receiver's mask. All final video shots, edits must be viewed in the TV-cutoff mode—switchable on most editing monitors to show what information will be in the final viewed product. See *overscan*.

cutting on the action: A production or editing technique in which two events are set in contrast to each other. As event "A" is taking place on the screen, the camera switches to event "B" before "A" has ended; often used to develop plot, create tension or produce contrast.

cutting on the reaction: A production or editing technique in which one event is followed by a scene that shows the results of that event or the impact of the event on the plot or characters.

cutting to tighten: An editing procedure used to shorten a series of shots. Used to eliminate excess footage and to produce a coherent whole.

cyc (cyclorama): Large "J" profiled background scenery, usually painted a solid color, eliminating any spatial frame of reference (limbo) in the scene's environment.

decibel (DB): A subjective evaluation of the volume of any particular sound in relation to other sounds; an evaluation of the strength of a signal in relation to a predetermined reference level.

decision point: A point in an interactive program where a user must make a decision, such as choosing an answer or deciding on an appropriate action.

degauss: To demagnetize—as in degaussing a CRT monitor.

demodulated: Description of a signal stripped of the carrier frequency onto which it was modulated; a signal is demodulated after it has been broadcast and prior to its display.

depth of field: The distance between the first object in focus and the last object in focus within a scene as viewed by a particular lens. Focal length of the lens and the F-stop used can affect the depth of field. Wider angle and smaller F-stop create greatest depth of field.

dew control: A warning system that indicates the presence of too much moisture for the safe operation of a video recorder.

dichroic conversion filter: A camera lens filter that balances the color values of objects in direct sunlight so they will match the value of scenes shot under artificial light.

dissolve: The slow crossfade; one picture gradually fades out as the next picture gradually fades in.

dolby noise reduction: An electronic device that reduces background noise and creates a better quality audio signal.

dolly: 1) Wheels on the feet of a tripod. 2) A wheeled camera-moving device that can run on track or rubber tires. 3) The act of moving the camera in a horizontal plane while a shot is in progress—usually forward and back.

DOS 3.3: The Apple Disk Operating System in its final 3.3 version. The operating system allows applications programs to be loaded from floppy disks and run. The original "DOS 3" was created in 1978 to support the Apple Disk II drive. DOS 3.3 works in Apple II+ and IIe machines.

downlink: Industry jargon for a satellite receiving dish, or for the process of beaming signals from satellites down to earth stations.

downstream keyer: A special-effects generator that enables the technical director (TD) to insert or key over composite video signal just before the video signal leaves the switcher to go over the air.

dropout: Loss of a portion of the video picture signal caused by lack of iron oxide on that portion of the videotape, or dirt or grease on that part of the tape.

dropout compensator: Circuitry that senses signal loss produced by drop-out and substitutes missing information with signal from the preceding line. If one line drops out of a picture, it is filled in with the preceding line, resulting in no drop-out on the screen.

dub: The duplication of a videotape. A rerecording process in which new dialogue in a foreign language is substituted for the original dialogue.

dubbing: Duplicating an audio or videotape. Dubbing puts the resulting copy one generation away from the original material or edited master.

earth station: An antenna that receives or sends satellite signals.

ECU: Extreme close-up.

edit code: Time code—videotape retrieval code added to the original recording and utilizing a time structure of hours, minutes, seconds and frames to locate a particular frame in the tape. Can be read off the screen through the use of a "window" burn-in over the screen.

editing deck: A specially constructed videotape recorder that has, in addition to the capability to record and playback, circuitry and controls that permit insert editing and/or assembly editing. Used in conjunction with a second source recorder containing elements to be edited into the master tape, which is on the editing deck.

effects buttons: The push-button controls on a special-effects generator (SEG) that indicate the special effects (inserts, wipes, keying, etc.) available on that SEG and that are engaged when a particular effect is desired.

effects channel: The bus in a three-bus switcher that is set aside to produce special effects.

eight-pin connector: A type of jack commonly used for the VTR-to-monitor connection.

electret condenser microphone: A very sensitive microphone requiring a DC power supply (a battery built into the microphone).

electronic editing: Repositioning the video signal segments on a reel of videotape without physically cutting the tape; a rerecording of the video signal elements in a different order. The edited version will be one generation away from the recordings from which it was assembled.

electronic news gathering (ENG): News gathering, usually employing a minimum of portable equipment to get the job done: camera, microphone, battery belt, sun gun battery light, video recorder (or Recam one-piece unit) and a crew of two.

electronic viewfinder: A small TV screen attached to the video camera that allows the operator to view the scene exactly as it is being viewed by the camera (if the finder has been properly tuned).

equalization (EQ): The normalization of an electronic signal, either audio or video. Adding EQ in audio means reshaping the frequency response to emphasize certain frequency ranges and eliminate others.

equalizing AMP: A video circuit that is preset to provide a certain equalization to the video signal.

erase head: Either an audio or video head that erases the signal on a videotape prior to the recording of a new signal on that tape. An electromagnet that neutralizes the data in the tape's oxide particles.

establishing shot: A long shot used to establish the environment in which the action to follow will take place.

external keying: A keying effect accomplished when a particular camera is assigned to supply the key signal through a special-effects generator (SEG) as opposed to internal keying, in which any of the cameras can supply the key signal. See *keying*.

extra: A player who appears in a program but has no lines of dialogue except where mob voices or voices in unison are required.

fade: To vary the strength of a signal as fading in or out—from black or to black.

fader: A sliding potentiometer (POT) control which diminishes the audio or video signal.

fiber optics: A technology that transmits voice, video and data by sending digital pulses of light through hair-thin strands of flexible glass.

field: The electronic signal corresponding to one passage over the raster area by the scanning spot; 262.5 lines (1/60th of a second) in the American system; half a frame. Two fields interlaced make one frame of video.

field blanking: Field retrace period—that period of time during which the field scanning spot returns from the bottom to the top of the raster area (the screen). Occupies about 15 to 20 lines of the 525-line system.

field frequency: The number of fields scanned per second; 60 fields are scanned per second in the American system.

field time base: The pattern of and points at which a field changes; 60 Hz is the field time base of the American TV system; the field time base must be kept steady and as constant as possible for the best picture. A time base corrector (TBC) maintains this stability and is part of most professional editing and recording systems.

fill: A segment at the end of a program that can be eliminated if necessary.

fill light: The illumination of shadowy areas in a scene to establish the proper brightness ratio or contrast ratio within the scene.

film chain: A special motion picture projector combined with a video camera to turn film into video.

film transfer: High-quality motion picture film made from an original videotape.

first generation: The original recording of a tape segment. The first time a signal is recorded on tape. Every subsequent recording of that segment will be one generation removed.

flag (French flag): Black cloth hung in front of a lamp to prevent stray light from flaring into the lens of the camera.

flagging: Picture distortion caused by improper operation of the VCR or VCR playback monitor timing coordination. Also caused by excessive copy-guard on commercial videocassette tapes.

flat: A unit of scenery, essentially the same as used in a stage setting. A movable wall.

flat light: Lighting a scene or setting with overall brightness without noticeable modeling or highlights.

floor manager: The production coordinator in charge of all floor operations not involving engineering. He or she supervises the erection of sets, placement of props, live sound effects, talent, cueing and the like. The floor manager is the director's representative in the studio or on the floor.

flowchart: A line diagram that represents the flow of information in a computer or interactive program.

fluid head tripod: A tripod whose camera mount consists of two metal plates. The upper, rotating plate rests on a bed of fluid and the movement provided is very smooth. Necessary for professional pans, tilts and other camera moves.

flyback: Retrace—the movement of a scanning spot from the end of a line or a field to the beginning of the next line or field.

flying spot scanner: Film-to-video transfer system utilizing an electronic shutter.

focus puller: Member of a video/film crew who is responsible for controlling focus of the lens while the camera or talent is moving.

fog filter: A lens filter that lends the effect of fog to a scene in increments of density from .05 to 3.

follow focus: The continual adjustment of the lens to keep an object in focus as either the object, camera or both are moving.

foot candle: 1) Amount of illumination received by a surface 1 foot from a lighted candle. 2) Also lumens per square foot. 3) The measurement of the intensity of light on an area measured with a light meter calibrated in foot candles. Example: The scene's brightness ranges from 50 to 200 foot candles.

format: As in recording format, 3/4-inch U-Matic and 1/2-inch VHS are two formats commonly used.

frame: A complete video picture composed of two fields. A total scanning of all 525 lines of the raster area, occurs every 1/30th of a second.

frame frequency: The number of frames occurring in a given period of time; usually 30 frames per second, or one half the field frequency of 60 Hz.

framestore: A device that records and stores video information that it can retrieve in the form of a still-frame picture; information is stored on a magnetic disk, or diskette.

frequency: The number of times a signal vibrates each second; expressed as cycles per second (CPS), or more usually as hertz (Hz).

frequency modulator: An electronic circuit that produces a carrier wave signal on which the audio or video signal is impressed.

frequency range: Frequency response; the width of frequencies from the highest to the lowest frequency that a piece of equipment is capable of handling without distortion.

fresnel lens: Fresnel spot; a specially constructed lens that produces a soft-edged concentration of light; used as a lens in a spotlight housing. (Pronounced "fren-nell.")

friction head tripod: Inexpensive tripod/camera mount that uses two plates pressing together, relying on friction to damp any fluctuation of the camera during a pan or tilt. Does not provide smooth camera movement.

F-stop: A calibrated control that indicates the amount of light passing through a lens to the target area. A linear, theoretical index, not relating to the actual amount of light transmitted by the lens. Actual transmission is measured in "T-stops," which are determined on an optical bench.

FX: Extraneous effects.

gaffer: Chief set electrician.

gain: Amount of signal amplification. Turning up the gain means increasing the strength of the signal. Used in both audio and video.

gap: The small space in an audio or video head across which the magnetic field is produced when recording and induced on playback. The audio and video heads are small, horseshoe-shaped electromagnets, and the gap is the space that tape must contact for good record playback.

gel (gelatin): Translucent celluloid light filter altering the characteristics of the light source.

genlock: Device synchronizing video signal sources. Circuitry that locks the sync generator, which is used to control cameras, and the special-effects generator (SEG) to the sync signal from a prerecorded tape on a VCR so the signal from that tape can be mixed through the SEG with live camera signals to create a composite image.

geosynchronous orbit: The altitude 22,300 miles above the equator at which a satellite's orbit is synchronized with the earth's rotation, making a communications satellite appear stationary.

glitch: Random video picture noise appearing as an ascending, horizontal bar across the screen.

gobo: A decorative frame through which the camera may shoot for special effects; a non-transparent black screen.

gray scale: The number of steps from black to white that a camera can resolve; can be used as an equivalent to the constant ratio; a test pattern.

grid: The cross-hatch of metal pipes for hanging lights in a video studio.

hand held: Using a camera without tripod or dolly.

hard disk: A mass storage computer device using an encapsulated rotating disk coated with magnetic medium and read by an electro-magnetic head. Hard disks hold far more data than floppy disks—in the megabyte (million bytes) range. Most modern business applications require hard disks for storage.

HDTV (high-definition television): Sharper video image than produced by the current 525-line scan technology. Doubles the number of scan lines to 1125 lines per second.

head: Audio or video; a small electromagnet that pulses magnetic signals onto a videotape moving past it or reads signals off a recorded tape; audio heads are usually stationary, while video heads move in reverse of the tape's direction in most VCRs.

head drum assembly: That portion of a VCR in which the video heads and their related mechanical and electronic controls are located.

headset: One- or two-way communications device used between the control room and the camera/floor personnel. Usually earphones and microphone in a single unit.

head shot: A close-up view of a subject.

heads out: A reel of tape is wound so that the beginning of the program is at the beginning of the tape; a rewound tape. Opposite of "tails out" tape where the end of the recorded session is at the beginning of the reel of tape. This is normally done to prevent magnetic bleed-through when a tape is stored for some time.

helical scan recording: A type of video recording (1-inch) in which the video heads and tape meet at such an angle that the resulting pattern on the tape is a long, diagonal series of tracks from the video heads. Each diagonal stripe contains the full information for one field of video picture. Named after the helical path the tape describes between supply and take-up reels.

hertz (Hz): The international electronic term for cycles per second.

high end: The highest frequency information of an audio or video signal. In audio circles, the high end refers to the treble portion of a signal.

highlight: An area of great brightness on a video display.

high resolution: Descriptive of a camera or monitor capable of displaying a great number of scan lines (1000 to 2000), which produce a very sharp and detailed picture. "Hi-res" monitor.

horizontal sync: The sync pulses that control the horizontal line-by-line scanning of the target area by the electron beam.

hue: A term used to describe the dominant wave length of a color in a range that runs from red to yellow, to green, to blue, to violet and back to red. All colors have a hue.

IATSE (International Alliance of Theatrical Stage Employees): Set workers. Many camera-men and other operators belong to this union, also known as IA.

IC (integrated circuit): Very small electronic component containing a photo-etched miniature circuit.

IEEE (Institute of Electrical and Electronic Engineers) scale: A waveform monitor scale in keeping with other IEEE standards and recommendations of video broadcasters and manufacturers.

image: The picture on the video screen, usually measured by its luminance values.

image plane: The point behind the lens at which the image collected by the lens is cast; in video, the image plane is the surface of the video tube target area.

image retention: Lag—the vidicon pickup tube's tendency to retain an image on its target area after it has stopped scanning that image. Extreme image retention results in the image being burned into the target area.

image transform: Proprietary computerized high-quality videotape-to-film transfer system.

impedance: The AC resistance of a component to the flow of a signal; expressed in high or low impedance.

inlay: Mortise-keyed insert; static matte insert. An insertion effect in which the fill signal is static and of a predetermined shape.

insert: A general-effects term meaning the introduction of a secondary signal into an already existing picture; accomplished by keying, wiping or crossfeeding.

insert edit: The insertion of a segment into an already recorded series of segments on a videotape; the inserted segment replaces one that must be the same length. Insert edits demand that the segment be cut in precisely, since already recorded information exists following the insert edit on the original tape.

insertion loss: The loss of signal strength that occurs when a piece of video or audio equipment is added to the path of the signal flow from origin to display; can be corrected by using an amplifier to build up the signal strength.

insert stage: Small studio for minor tabletop or closeup videotaping.

interactive video: Any video program in which the sequence and selection of messages is determined by the user's response to the material.

intercom line: Usually the audio connection between the video director and the members of the crew.

intercutting: A production technique in which a cut is made from a scene (long shot) to a detail of that scene (close-up) to clarify or emphasize a point.

interface: A device that allows devices to communicate with each other. An interactive interface translates the language of a computer into signals a videodisc player can interpret.

interlace: A scanning method in which the lines of two fields are combined into a frame in such a way that all the lines of each field are visible as part of the frame; the positioning of 262.5 lines from one field with 262.5 lines from the next field to form a full 525-line frame.

internal keying: A method of keying in which the key signal can be sent through the SEG from anyone of the cameras already in use. See *external keying.*

inverter: DC- to AC-current converter, similar to a rectifier.

IPS (inches per second): The customary way of measuring tape speed on an audio or video recorder.

iris: Iris diaphragm—metal leaves that create an aperture used to control the amount of light passing through the lens. Iris openings are measured in F-stops.

ITFS (Instructional Television Fixed Service): Group of TV channels in the UHF frequency range set aside for educational use.

jack: Plug-in electronic connection or connector.

jack bay: See *patch panel.*

jacket: The protective and insulating cover of a cable.

jitter: A tendency toward lack of synchronization of the picture. It may refer to individual lines in the picture or to the entire field of view.

joystick: Control stick for operation of electronic equipment such as editing controllers and video switchers—SEGs.

jump cut: Bad or jagged edit of tape or film. Jump of action within a scene that does not take advantage of a cut-away to conceal the slight change of position of the subject caused by editing out some of the scene's information.

junction box: Portable terminal box for AC power; also portable terminal for cable interfacing.

Kelvin: Kelvin temperature—used for measurement of light source color temperature. Stage lighting is based around 3200° Kelvin, sunlight is approximately 5800° Kelvin, while high-overcast, bright days are close to 6000° Kelvin. If Kelvin temperatures are changed in a scene, the whites must be re-balanced (camera's white balance control button) for accurate rendition of color. The lower the Kelvin, the warmer the light color.

keying: Keyed insert, inlay insert. One video signal being controlled by the waveform of a second video signal when they are combined to form a composite picture. The signal from source 1 fills in the scanning lines of the total picture of source 2 at the points where the picture goes above a certain, pre-established gray level.

keylight: The spotlight or main light on a scene which emphasizes the important objects in that scene.

kilohertz (kHz): A unit of frequency equal to 1000 Hz.

lag: "Ghost" image retained when fast action occurs in the presence of insufficient illumination.

lap: Lap dissolve—a cross dissolve into new material while dissolving out old material.

leader: Audio or video countdown introduction at head of tape for identification purposes. Also used as cue leader, is often used at the head and end of tape, or between sections of raw, unedited material.

lens speed: Measurement of the ability of a lens to collect light—usually expressed by its lowest F-stop number.

level: Audio or video amplitude or intensity. Also as in, "Give me a level," a test of same.

light emitting diode (LED): A brightly lighted semi-conductor component.

lighting ratio: The brightness level of the fill light compared to the brightness level of the key light, or the shadowy areas compared to the brightly lit areas measured as a ratio determined by the F-stop of the lens. A 1.2 ratio means that the key is one F-stop brighter than the fill; 1:3, a stop and a half; and 1:4, two stops.

light level: The intensitey of light available measured in foot candles or lux (10.75 lux = 1 foot candle).

limiter: A circuit that shapes a signal sent through it to conform to a certain preset tolerance; used to limit distortion in audio or video.

linear video: Any video program in which the structure, sequence, pace and selection of material is predetermined and invariable (as opposed to "interactive video").

line frequency: The number of lines scanned in one second; in the U.S. system it is 525 x 60, or a line frequency of 15.7 kHz.

line matching transformer: An audio device used to match the impedance of a microphone to the input impedance of a mixer, VCR or amplifier; a device that changes the output impedance of a microphone from low to high, or vice-versa.

line out: See *video out.*

line period: The length of time it takes for a line to be scanned and then retraced to the point where scanning of the next line will begin.

line scanning: The path over the target area of the electron beam, as it moves from the left edge across the area.

line time base: The control of the horizontal deflection of the scanning spot so that it starts to scan each new line at exactly the right moment.

load: To place a termination across a video or audio line.

long shot: A camera angle of view taken at a distance and including a great deal of the area scene.

low light lag: A blurring, image retention effect that occurs when a vidicon tube is operating in insufficient light.

low pass filter: A filter, often used on two-way cable systems, that inhibits the flow of high-frequency information along a cable while it allows low-frequency information to pass.

low power television (LPTV): Authorized by the FCC for broadcasting to small geographic locations. LPTV stations can be "squeezed" between existing channels without creating interference.

lumen: A measure of light quantity.

luminance signal: The black-to-white brightness values of a scene that produce the black-and-white display image.

lux: The metric measurement of light quantity taken at the surface where the light source is illuminating; one footcandle equals 10.76 lux.

machine-to-machine edit: Transfering video and audio material from one recorder to player to another recorder, in which the edit is made by a simple assemble edit button. A glitch or distortion may show at the edit point. Crash edit.

macro lens: A magnifying lens capable of focusing down to a few inches.

magicam: Proprietary matting system for using small-scale model sets and matting actors electronically to save set construction costs.

mark: Gaffer tape indication on studio floor to indicate a camera or talent position.

master: Original completed videotape. Derivatives: "B" master, dubbing master, editing master, submaster.

master VTR: When duplicating tapes, the machine that plays the "master" tape being duplicated.

match cut: Editing another camera view of an identical moment in the recorded action.

match dissolve: Fading to, or dissolving to, an identical camera position.

matching transformer: A circuit that changes the impedance of a video signal, often from 75 to 300 OHMs. For audio, line matching transformer.

matte: A film term sometimes used in video work to denote a keyed effect, an insert of video signal information keyed from one source into a second video signal.

medium shot: Camera angle of view between close-up and long shot. A view of the head and shoulders of a subject, as opposed to the head only (close-up), or full body (long shot).

Mickey Mouse: Using obvious primary colors or standard gray scale shades to do titles or backgrounds: "Mickey Mousing it."

microsecond: One millionth of a second.

minicam: Lightweight, often self-contained, portable ENG-type camera.

mini plug: A female receptacle that accepts a mini-jack; similar to a phonoplug in design, but much smaller.

mistracking: Improper tape path and tape-to-head contact resulting in bursts of noise appearing in the picture.

mix: Session in the recording studio. Usually refers to an audio session where diverse audio elements are combined on single or parallel sound tracks.

mode: Electronic setting activating specific circuits in a system, i.e., record mode, playback mode.

modulation: The process of adding audio or video signals to a predetermined carrier signal.

moire: Optical disturbance caused by interference of similar frequencies.

monaural: Single sound source to both ears.

monitor: TV set without receiving circuitry used to directly display the composite signal received from a camera, videotape or special-effects generator.

montage: Visual blending of several scenes.

mount: Refers to lens mounting on a camera, as in "C" mount, which accepts type C screw-in lenses.

MS-DOS: An operating system developed by Microsoft. A 16-bit system using microprocessors in the 8086-8088 family (including the new 80186 and 80286 processors). It is considered the industry standard at this writing but will be replaced by the new OS/2 system.

multi-box: Generally a four-way electrical junction box used on location and in the studio.

multiplex: Signal conductor transmitting several different picture sources.

multiplexer: An optical system allowing a number of film and slide sources to feed video information into the same camera.

NAB: National Association of Broadcasters: standard-setting and fraternal organization of the broadcast industry.

NABET: National Association of Broadcast Employees and Technicians—a broadcast union.

NAEB: National Association of Educational Broadcasters.

needle drop: Single usage of a licensed or copyrighted piece of music.

nemo: A remote pick-up.

noise: Any unwanted signal present in the total signal. Both an audio and video term to describe one signal interfering with another. Visual picture static.

non-composite video signal: A video signal containing picture and blanking information, but no sync signals.

NTSC: National Television Standards Committee—a broadcast advisory group established in the 1940s that recommended standards to the FCC for the 525-line, 60-field system.

offline: Using the offline editing suite with 1/2-inch video recorders to rough cut the 1/2-inch transfers of the master tape footage. An off-line

1/2-inch tape is produced and an editing decision list is created. The off-line rough cut can be used for client approval, and the edit list is used to speed editing of the final cut during "on-line" editing when special effects, full sound mix and all dissolves are added to create the edited master tape.

offline tape transfers: Original camera footage is transferred to 1/2-inch or 3/4-inch tape copies in order to be edited into an offline copy (rough cut) for client approval.

off-mike: Not within pick-up pattern of the microphone.

omni-directional: Microphone with a pick-up pattern that is generally uniform in all directions.

online: Making use of the editing suite for final assembly of the videotape, sound mix and special effects. Opposite of *offline*.

open MIC: Live microphone.

operating system: A set of computer instructions devoted to the operation of the computer itself. An operating system must be present in the computer before any applications programs can be run. Applications programs are designed to run with specific operating systems. The Apple's operating system is ProDOS, or DOS 3.3. IBM's operating system is PC-DOS (a version of MS-DOS).

optical videodisc: Video playback system in which a low-power laser is reflected against a disc's microscopic pits to retrieve frames of prerecorded information. A disc contains as many as 54,000 individual frames, any one of which can be located and displayed almost instantly.

OS/2: current IBM operating system (replacing MS-DOS).

output: The terminal point of a unit of electronic equipment from which the signal is taken.

overlay: The ability of some external computers to superimpose text or graphics over a video image, creating a picture that combines both video and graphics.

overlay insertion: Self-keyed insertions, moving matte insertions. An insertion effect in which the fill signal is a moving object that determines its own parameters as it moves.

overscan: Video picture beyond the area of normal screen size.

pad: Cushion or fill.

paint pots: Controls of a colorizer for mixing colors electronically. Rheostats.

PAL (phase alternate line): 625-line, 50-field system used in the U.K., Western Europe, Scandinavia, Australia and South Africa. (U.S. uses NTSC 525-line, 60-field system, which is somewhat less sharp than PAL.)

pan: To follow action by swinging the camera left or right.

patch: A set of computer instructions that allows a computer to run a program that was not originally designed to run on that computer. The instructions are usually typed in or loaded into the computer's operating system before the program is loaded from disk. Example: To allow a Franklin computer to run Apple's ProDOS operating system, you must first type, "265B: EA EA (return) 2000G (return)" to "patch" ProDOS into the Franklin system.

patch cord: Any cable with a jack at each end used to connect audio or video components to each other.

patch panel: A plate with a number of female receptacles, each the termination of a different audio or video signal. Used with patch cords to make secure, but temporary, connections between components.

peak-to-peak voltage: The total voltage produced by a signal, determined by adding together the positive and negative extremes to which the voltage modulates.

peak white: The brightest, whitest portion of the picture signal corresponding to the highest level the signal attains.

pedestal: Black level. The minimum level that the blackest portions of the displayed signal are allowed to reach. To "lower the pedestal" means to lower the black level.

phase: The relative timing of a signal in relation to another signal; if both signals occur at the same instant, they are said to be in phase, at different instants—out of phase.

phone plug: Variety of jack often used as a microphone connector.

phono plug: Variety of jack most often used with audio amplifiers. Also known as an RCA plug.

phosphor: A chemical coating used on the inside of a cathode ray display screen (video screen). When hit by electrons it glows. A narrow beam of electrons causes a pinpoint to glow. As the beam scans, an image is "painted" on the inside of the tube.

picture lock-up: Synchronizing the picture signal. Sync controls on a picture.

picture signal: The picture information part of the composite video signal; the portion of the signal above the pedestal.

pinch roller: A rubber roller in the videotape path that holds the tape against the capstan. The capstan spins, and the tape is pulled through the recorder.

playback: Function that induces the magnetic patterns on a videotape from that tape and into the recorder's circuitry in order to reconstruct the composite video signal for display.

plumbicon: Trade name of a special lead oxide tube that is more sensitive than a vidicon or saticon tube. Used in color cameras.

plus diopter: Lens accessory that fits over the camera lens to make the lens capable of extreme close-ups.

polarizing filter: Lens filter with polarizing properties that reduce the amount of reflected light coming into the lens.

pop filter: Spongy rubber cap placed over a microphone to reduce breathy "pops" from the letter "P" or "B."

pot: Potentiometer—variable resistor used to control volumes/levels.

pre-production: Covers all activity prior to actual taped production.

preview: The monitoring of a video signal prior to its being processed through a special-effects generator (SEG).

preview bus: A row of video source buttons that enables the technical director to look at any video image before it goes out onto the air or onto tape.

processing amplifier: Proc-amp, signal processor, helical scan processor. A unit inserted in the line between any two components through which a composite video signal travels; serves to stabilize the composite signal, regenerate the control pulses and, in certain models, change the gain and pedestal to improve contrast.

program bus: The master bus on a switcher, which controls the output signal of the switcher.

projection television: A combination of lenses and mirrors that projects an enlarged TV image on a screen.

pull back: A dollying back from a subject with the camera as opposed to zooming back.

pull focus: To intentionally go from soft focus to sharp focus, or vice versa.

pulse: The variation of a constant signal for a certain period of time.

pulse distribution amp: Amplifier designed to boost signal strength of control signals to a number of cameras, SEGS, etc.

punch up: To engage a function button, as in punching up a special effect on the SEG.

push rod: Rod with handle that permits control of focus and zoom from the back of a video camera.

quad: Quadruplex—four head recording system, such as 2-inch (51mm) quad, which writes information on successive vertical stripes (1,2,3,4 1,2,3,4), etc. The video read-write heads pass the tape at an angle perpendicular to its path.

random access: Simple retrieving of stored magnetic information regardless of where it is located on the tape or videodisc system.

raster: The pattern described by the scanning spot of the electron beam as it scans the target area of a cathode ray tube, the pattern of scanning in both the pick-up and display tubes.

raw stock: Unrecorded videotape.

real time: Original time span without compression or selective grouping.

registration: An adjustment associated with color TV to ensure that the electron beams of the three primary colors of the phosphor screen are hitting the proper color dots/stripes. A similar adjustment of the color tubes on a camera.

resolution: A subjective evaluation of the amount of detail in a picture.

RF adaptor: RF amplifier; RF modulator-convertor. A unit that accepts the composite video signal to modulate a carrier frequency and produce a broadcast signal on a standard TV channel.

RGB signal: The chrominance information—red, green and blue.

roll: Loss of vertical sync, causing the picture to move up or down the screen.

room tone: Recorded ambient noise used in sound tracks. Can be laid under studio recorded voices to simulate presence in the room.

rotary erase head: Set of heads on the rotating video head assembly that erase the video signal during recording and editing; usually positioned one scan line in front of the video heads; produces cleaner edits than a stationary erase head.

rotary idler: Stationary guide along the tape path.

rough edit: A rapid assembly of various segments in the order they will appear in the final program; not a finished tape.

safe area: Ninety percent of the video screen, from the center of the screen; that area of the screen that will reproduce on any TV screen.

safe title area: Eighty percent of the video screen, from the center of the screen, which will produce legible titles and credits on any TV set.

safety: Extra copy of a master videotape made as a back-up to the final, edited program.

scanning: The action of the electron beam as it traces a pattern over the target area of the camera pick-up tube in order to convert the light values of each spot on the area to a corresponding electrical signal.

scoop: 500-watt circular floodlight, used for area coverage.

SECAM (sequential couleur a memoire): Sequential color and memory. A color video system developed by the French, which differs greatly from both the PAL and NTSC color systems. Used in France, USSR and Eastern Europe. 625 lines, 50 fields.

SEG (special-effects generator): For video effects, such as wipes, fades and dissolves.

shadow mask color tube: Dot matrix color picture tube; a color tube equipped with a sheet metal frame with half a million holes in it. The sheet metal (which is the shadow mask) is placed between the electron guns, which beam the picture signal, and the phosphor-coated screen. Improves picture contrast.

signal path: The movement of the signal from the point of origin to the point of display through a component or series of components.

signal-to-noise ratio: The higher the signal-to-noise ratio (the more signal, the less noise), the better the quality of the resulting sound or video picture.

skew: The tape tension between supply reel and the first rotary idler of the tape path around the head assembly of a VCR. Skew must be properly maintained or picture instability will result.

slate: To slate is to identify a scene. A clap-board with scene information on it inserted at the beginning of the scene. Also an electronic slate.

slow-mo: To slow down or vary the tape speed below normal. Also name for a special videodisc used for sports recording. Slow-motion.

SMPTE: Society of Motion Picture Technicians and Engineers.

snow: Random noise on the display screen, usually resulting from dirty record/play heads.

soft key: A chroma key in which the edges of the insert image blend with the background. Permits keying transparent objects and shadows into the background picture for a more realistic key. Also called luminance keying.

special-effects generator: A unit used in video production to mix, switch, and otherwise process various video signals to create a final signal known as the program signal.

speed: Called when the VTR and camera have reached full recording speed—before "action" is called.

spillover: The leakage of one signal from one line to an adjacent line or from one tape layer to another.

split screen: A special effect utilizing two or more cameras so two or more images appear on the same screen. Also used in post-production to produce the same effect from two separate prerecorded images.

spyder: Small camera dolly.

start mark: Sync indication in either audio or the video track marking the head of the track.

step-down transformer: An electronic circuit that can change electric current from one voltage to another; the most common is 220V to 120V.

steps: Term used to describe the number of controls on a colorizer; the control for each color is called a step.

still frame: An individual frame of video being held as a continuous shot.

subcarrier frequency: The frequency on which color information is modulated. In the U.S., it's 3.58 MHz.

super: The superimposition of one video image over another, using the fader controls of the special-effects generator.

super cardioid: A microphone with a very directional pick-up—picking up sound coming at the front of the microphone only.

switcher: A unit that allows the operator to switch between video camera signals. In a broader sense, refers to the unit responsible for special effects and controlling what images are sent to and from what sources.

switcher-fader: Switcher with an added device that permits fading from one video signal to another or to superimpose two images.

sync: Synchronize: various drive pulses, both horizontal and vertical, which maintain the horizontal and vertical scanning procedures of the video picture from camera to display.

sync generator: A pulse generator that produces the sync signals necessary to integrate the functioning of various pieces of video equipment in relation to each other and the video signal.

sync mark: Editor's reference mark.

tally: A system of audio intercommunications among various members of the video production crew. A tally light is on top of each camera and glows when that camera is on.

tape path: The circuit the tape runs from supply to take-up reel past the erase head, video heads, audio/control track head and between capstan and pinch roller.

target area: The face of the vidicon tube or other camera CRT pick-up tube. This area (opposite the cathode heater) is where the image formed by the lens is transformed into an electronic signal. On the outside face of the tube, the image from the lens is focused; and on the inside, the image is read by the electronic scanning beam. A circuit is completed at each point the beam strikes the target, and because voltage is being applied to the target area, a certain resistance results, which gives a voltage variation or video signal.

tearing: Occurs when horizontal sync is lost or distorted in a picture, resulting in some of the horizontal lines being out of place.

tension: The pull of the capstan assembly on the videotape to keep it against the video head drum assembly.

three-tube color camera: A color-capable camera that produces a color signal through the use of three pickup tubes, each assigned to one of the primary colors.

tilt: To move the camera up toward the ceiling (tilt up) or down (tilt down).

time code: An identifying code used to measure and locate specific frames on the videotape. The readout is shown in hours, minutes, seconds and frames. Time code can be logged during taping by a camera assistant or the tape engineer. Time code is normally burned into a "window" on off-line VHS 1/2-inch copies for referencing frames during the off-line edit. Time code is the basis for locating scene in and out points during editing.

timed video still: A frame of information left on the screen for a specific length of time.

title crawl: A roller-like device used to roll credits or titles across the video screen. Mechanical and electronic crawls are available.

tongue: Dolly-mounted camera boom.

touch screen: A user response device that allows the user to input information into an interactive program by touching a particular area of the video monitor's screen. The touch either breaks a field of intersecting beams or compresses two membranes covering the screen to produce an intersecting of X and Y axes, which have meaning in the computer program that runs the interactivity.

tracking: The angle and speed at which the tape passes the video heads.

trucking: Moving the camera left or right on a tripod with a dolly.

turnkey: A system that is provided by a manufacturer in a single package. The Sony model 2000 intelligent video system is a *turnkey* system in that the user is supplied with the computer, videodisc player and all necessary hardware and software to create and interpret interactive videodisc programs in a Level 3 environment.

TV storyboard: Sheets of paper with blank TV screens on them used for sketching out the action of the program.

tweak: To exactly align electronic equipment.

two-shot: Video picture showing two performers or two objects of major interest.

2:1 interlace: Scanning system in which the horizontal and vertical control pulses are locked together so that they occur at the correct time in relation to each other.

type B format: Helical record/reproduction format using segmented scanning process—two or more video heads used to divide or segment the video information as it is written on the videotape.

type C format: Helical VTR record/reproduction characterized by unsegmented scanning process. (See *type B format*.)

uni-directional: Microphone pick-up pattern that accepts only sound coming in front of it.

varied outcomes: The situation that results when the different choices at a decision point branch to very different results. Used in interactive video to show the consequences of different decisions.

vector scope: Round (green) oscilloscope to align amplitude and phase of the three color signals (RGB).

vertical blanking: Field blanking; the blanking of a signal during scanning; when the scanning spot is flying back from scanning one field to begin scanning the next field, and at which time blanking and sync pulses are introduced to the signal.

vertical interval: Moment measured in microseconds during which the electron scanning beam returns to the top of the video tube.

vertical scanning: The field-by-field scanning of a picture at the rate of 60 fields per second (in U.S., Canada, Mexico and Japan).

vertical sync: The sync pulses that control the vertical field-by-field scanning of the target area by the electron beam.

video: Television and the technical equipment and events involved in creating television; the visual portion of a signal containing both sight and sound information; an alternative to broadcast television. (Latin: "I see.")

video amplifier: A circuit that can increase the strength of a video signal passing through it.

videodisc: An electronic communication medium that has audio and video images encoded in grooves on a flat surface. The grooves are read by an optical laser and transformed into FM signals, becoming sound and picture information. Used in home entertainment and interactive video training programs, point-of-purchase application and multiple choice activities where the user responds to prompts and controls the random access of information on the disc.

video distribution amplifier: A special amplifier for strengthening the video signal so it can be supplied to a number of video monitors at the same time.

video frequency (VF): A composite video signal unmodulated by radio carrier frequency.

video gain: The amplitude of the video signal; the control on a VCR that determines the "volume" of the video signal.

videography: Term used to describe videotaping in a photographic sense; i.e., making a motion picture with video equipment.

video in: Line in—jack through which a video signal is fed into any component.

video out: Line out—jack from which a video signal is fed out of a given component.

video waveform: The pictorial display on a special oscilloscope of the various components of the video signal, used to check the integrity of the signal and signal components.

voice-over: A voice speaking over action; narration.

volume unit meter: VU meter. Meter for monitoring the amplitude of audio or video signals.

waveform monitor: Special oscilloscope used to display the video waveform.

white level set: White balance. A camera control that establishes the luminance level for a color camera.

wide open: Description of a lens at its widest lens stop in order to shoot under low-light levels.

wild footage: Audiotape recorded out of sync with any particular video picture for use in post-production as an audio track.

wipe: Term used to describe the SEG effect of replacing a portion of video signal A with video signal B. Also to erase a tape.

wrap: "That's a wrap"—to finish an entire shoot and pack up the equipment.

zoom lens: A lens with a variable focal length.

zoom shot: A camera movement that involves zooming in or out while the camera is live.

APPENDIX A: INTERACTIVE SYSTEM DEVELOPERS

ACCESS Network
295 Midpark Way SE
Calgary ALB T2X 2A8
CANADA
(403) 256-1100

Acorn Company
23 E. 10th St., Suite 414
New York NY 10003
(212) 673-3333

Action Systems Inc.
Suite 115
13747 Montifort Drive
Dallas TX 75240
(214) 385-0680

Advanced Systems Inc.
155 East Algonquin Road
Arlington Heights IL 60005
(312) 981-1500

Advanced Touch Systems
9669 Distribution Avenue
San Diego CA 92121
(619) 693-9001

AIMtech Corporation
77 Northeastern Boulevard
Nashua NH 03062
(603) 883-0220

Alamo Learning Systems
Suite 500
1850 Mount Diablo Boulevard
Walnut Creek CA 94596
(415) 930-8520

ALIVE Center
400 Wabash Avenue
Akron OH 44307
(216) 384-6413

Allen Communications
140 Wayside Plaza II
5225 Wiley Post Way
Salt Lake City UT 84116
(801) 537-7800

American Industrial Publications
PO Box 2831
Durham NC 27705
(919) 286-2199

APh Technological Consulting
55 North St. John Avenue
Pasadena CA 91103
(818) 796 0331

Applied Data Research
CN-8
Route 206 and Orchard Road
Princeton NJ 08540
(201) 874-9000

**Ashton Interactive Training
    Corporation**
PO Box 5619
Vestal NY 13851
(607) 748-4015

AST Research Inc.
2121 Alton Avenue
Irvine CA 92714
(714) 853-1333

ASVS Computer Systems
104 Viewcrest
Bellingham WA 98225
(206) 734-2553

AT&T Graphic Software Labs
10291 North Meridian—Suite 275
Suite 29-A
Indianapolis IN 46290
(317) 844-4364

AVAS
196 Holt Street—PO Box 1070
Hackensack NJ 07602
(201) 487-6291

Bank Street College of Education
610 West 112th Street
New York NY 10025
(212) 675-8566

Barco Industries Inc.
2211-B Executive Street
Charlotte NC 28208
(704) 392-9371

BCD Associates Inc.
205 Broadway Technical Center
7510 North Broadway Extension
Oklahoma City OK 73116
(405) 843-4574

Boeing Computer Services
PO Box 24346
Seattle WA 98124
(206) 644-6183

British Broadcasting Corporation
Room CG05—Woodlands—80 Wood Lane
London ENG W12 0TT U.K.
(01)576-0339

ByVideo Inc.
225 Humboldt Court
Sunnyvale CA 94051
(408) 747-1101

Carroll Touch Corporation
PO Box 1309
Round Rock TX 78680
(512) 244-3500

CDEX Intellisance Corporation
1885 Lundy Avenue
San Jose CA 95131
(408) 263-0430

CEIT Systems
Suite 202
25 East Trimble Road
San Jose CA 95131
(408) 943-9797

Celebrations Video
Suite 247
2509 North Campbell
Tucson AZ 85719
(602) 623-1400

Cincinnati Milacron
Department 85-E
4701 Marburg Avenue
Cincinnati OH 45209
(513) 841-7257

Compact Disc-Video (CD-V)
Coordinating Office
Parker Plaza
400 Kelby Street
Fort Lee NJ 07024
(201) 461-9874

Computer Sciences Corporation
Suite 100
11836 Fishing Point Drive
Newport News VA 23606
(804) 873-1024

Computer Systems Research Inc.
PO Box 45
Avon CT 06001
(800) 922-1190

Computer Teaching Corporation
Illini Plaza
1713 South Neil Street
Champaign IL 61820
(217) 352-6363

Comsell Inc.
500 Tech Parkway
Atlanta GA 30313
(404) 872-2500

Comware
4225 Malsbary Road
Cincinnati OH 45242
(513) 791-4224

Control Data Corporation
HGW-038
PO Box Zero
Minneapolis MN 55440
(800) 233-8908

Creative Technologies
  Corporation
4820 North Spring Street
Evansville IN 47711
(812) 422-4112

CV Mosby Company
11830 Westline Industrial Drive
St. Louis MO 63146
(800) 325-4177

CW Cameron Ltd.
Kirkhill House—Broom Road East
Newton Mearns—Glasgow SCO G77 5LL
UNITED KINGDOM
(041) 639-2000

Datamed
5202 Valmar
Concord CA 94521
(415) 798-6736

Data Processing Resources
PO Box 47944
Tulsa OK 74147
(918) 492-8873

Deltak Training Corporation
1751 West Diehl Road
Naperville IL 60566
(312) 369-3000

Destron
Suite 1902
180 North LaSalle Street
Chicago IL 60601
(312) 332-6800

Development Dimensions International
PO Box 13379
1225 Washington Pike
Pittsburgh PA 15243
(412) 257-0600

Diaquest Inc.
1442 San Pablo Avenue
Berkeley CA 94702
(415) 527-7700

Diebold Inc.
5995 Mayfair Road
North Canton OH 44720
(216) 497-4580

Digital Controls
6576A I-85 Interstate Court
Norcross GA 30092
(404) 441-3332

Digital Equipment Corporation
30 North Avenue
Burlington MA 01803
(617) 276-1431

Digital Research Inc.
Box DRI
Monterey CA 93942
(408) 649-3896

Digital Techniques Inc.
10 B Street
Burlington MA 01803
(617) 273-3495

Digital Videodisc Technologies Inc.
1704 Moon Avenue NE
Albuquerque NM 87112
(505) 292-1212

Discus Electronic Training
1160 Pittsford-Victor Road
Pittsford NY 14534
(716) 248-4301

Discworks
18 Ellery Street
Cambridge MA 02138
(617) 491-7731

DKW Systems Inc.
703, 9919—105 Street
Edmonton ALB T5K 1B1
CANADA
(403) 426-1551

Ear Three Systems Manufacturing Company
PO Box 9424
Arlington VA 22209
(703) 525-8770

Edudisc
1400 Tyne Boulevard
Nashville TN 37215
(615) 269-9508

Effective Training Inc.
9544 West Pico Boulevard
Los Angeles CA 90035
(213) 359-2603

Electronic Information Systems Inc.
Suite 650
5 Triad Center
Salt Lake City UT 84180
(619) 231-7774

Electronistore Services Inc.
Suite 360
2122 York Road
Oak Brook IL 60521
(312) 574-4911

Elographics Corporation
105 Randolph Road
Oak Ridge TN 37830
(615) 482-4100

Executive Technology Data Systems
Suite 200
34405 12 Mile Road
Farmington Hills MI 48018
(313) 553-6665

Flight Training Devices
PO Box 91723
Anchorage AK 99509
(907) 276-6719

**Ford Aerospace
Corporation**
10800 Parkridge Boulevard
Reston VA 22091
(703) 620-6800

General Electric Company
3135 Easton Turnpike
Fairfield CT 06431
(203) 373-3468

Gentech Corporation
4102 N St. Joseph Ave.
Evansville IN 47112

Global Information Systems
1800 Woodfield Drive
Savoy IL 61874
(217) 352-1165

Gould Inc.
Shawsheen Station
PO Box 3083
Andover MA 01810
(617) 475-4700 x9289

Health EduTech Inc.
Suite 350
7801 East Bush Lake Road
Minneapolis MN 55435
(612) 831-6246

Hitachi Sales Corporation of America
401 West Artesia Boulevard
Compton CA 90220
(213) 537-8383

Homecom Learning Systems Inc.
150 Bloor Street West—Suite 307
Toronto ONT M5S 2X9 CANADA
(416) 968-7155

IBM Corporation
PO Box 2150
Atlanta GA 30055
(404) 238-4250

IEV Corporation
3030 South Main Street
Salt Lake City UT 84115
(801) 466-9093

IIAT
(International Institute of Applied
    Technology Inc.)
14620 Southlawn Lane
Rockville MD 20850
(301) 424-7184

IMSATT Corporation
500 North Washington Street
Falls Church VA 22046
(703) 533-7500

Industrial Training Corporation
13515 Dulles Technology Drive
Herndon VA 22071
(703) 471-1414

Industrial Training Systems Corporation
20 West Stow Road
Marlton NJ 08053
(609) 983-7300

Info-Disc Corporation
Suite 134
4 Professional Drive
Gaithersburg MD 20879
(301) 948-2300

Information Design Inc.
Suite 3-3
145 Durham Road
Madison CT 06443
(800) 336-9273

Instructional Design International
1775 Church Street NW
Washington DC 20036
(202) 332-5353

Interac Corporation
Suite 175
12555 West Jefferson Boulevard
Los Angeles CA 90066
(213) 301-7640

Interaction Systems
130 Lincoln Street
Brighton MA 02135
(617) 789-5900

Interactive Performance Systems
PO Box 728
Kenosha WI 53141
(414) 658-2110

Interactive Technologies Corporation
Suite 315
9625 Black Mountain Road
San Diego CA 92126
(619) 693-1020

Interactive Technology Inc.
PO Box 948
Springdale AR 72765
(501) 442-0301

Interactive Training Systems
9 Oak Park Drive
Bedford MA 01730
(617) 271-0500

Interactive Video Concept Inc.
Suite 105, The Wilford Bldg
101 N. 33rd St.
Philadelphia PA 19104
(215) 387-0709

InterDigital Inc.
Water Street
Lebanon NJ 08833
(201) 832-2463

ISC Technologies Inc.
3700 Electronics Way
PO Box 3040
Lancaster PA 17604
(717) 684-9607

ITW Entrex
6615 West Irving Park Road
Chicago IL 60634
(312) 282-9440

JAM Inc.
300 Main Street
East Rochester NY 14445
(716) 385-6740

JVC (Victor Company of Japan Ltd.)
Los Angeles Liaison Office—Suite 500
6363 Sunset Boulevard
Hollywood CA 90028
(213) 059-0584

Kaset Inc.
14003 North Dale Mabry
Tampa FL 33618
(813) 962-7830

Kepner-Tregoe Inc.
PO Box 704
Research Road
Princeton NJ 08542
(609) 921-2806

Kinton
5707 Seminary Road
Bailey's Crossroads VA 22041
(703) 820-1000

Lang Learning Systems
Excelsiorlaan 23
B-1930 Zaventem
BELGIUM
(02)720-5006

LaserData Inc.
10 Technology Drive
Lowell MA 01851
(617) 494-4900

Learncom
Suite 2A
215 First Street
Cambridge MA 02142
(617) 576-3100

Learning Dynamics Inc.
PO Box 323
Needham MA 02192
(617) 332-7070

Learning Resources Inc.
PO Box 3416
Durham NC 27702
(919) 683-8050

Maritz Communications Company
Suite 1700
600 Renaissance Center
Detroit MI 48243
(313) 882-9100

Maryland Interactive Technologies
11767 Bonita Avenue
Owings Mills MD 21117
(301) 337-4117

Matrox Electronic Systems
1055 St. Regis Boulevard
Dorval QUE H9P 2T4
CANADA
(514) 685-2630

McDonnell Douglas Electronics Company
PO Box 426
St. Charles MO 63302
(314) 925-4972

McGraw Hill Book Company
28th Floor
1221 Avenue of the Americas
New York NY 10020
(212) 512-6589

MicroEd
PO Box 444005
Eden Prairie MN 55344
(612) 944-8750

Micro Touch Systems Inc.
10 State Street
Woburn MA 01801
(617) 935-0080

Minnesota Educational Computing Corp.
3490 Lexington Avenue North
St. Paul MN 55126
(612) 481-3670

Mitsubishi Electric Sales
Box 6007—CRL
5757 Plaza Drive
Cypress CA 90630
(714) 220-4723

MWEI Teaching Technology
PO Box 3808
San Luis Obispo CA 93403
(805) 541-3100

North American Phillips Corporation
1111 Northshore Drive—Box 204
Knoxville TN 37919
(615) 558-5110

NCR Corporation
Building USG-1
Dayton OH 45479
(513) 445-5251

Network Data Systems
1535 East Pierson Road
PO Box 663
Flushing MI 48433
(313) 659-4045

New Media Graphics Corporation
Suite 5
279 Cambridge Street
Burlington MA 01803
(617) 272-8844

Nissei Sangyo America Ltd.
Suite 401
1701 Golf Road
Rolling Meadows IL 60008
(312) 364-2463

OmniCom Associates
407 Coddington Road
Ithaca NY 14850
(607) 277-0405

Online Computer Systems
20251 Century Boulevard
Germantown MD 20874
(301) 428-3700

Optical Data Corporation
PO Box 97
66 Hanover Road
Florham Park NJ 07932
(201) 377-0302

Panasonic Industrial Company
One Panasonic Way
Secaucus NJ 07094
(201) 392-4602

Parallax Graphics
1095 East Duane Avenue
Sunnyvale CA 94086
(408) 720-1600

Perceptronics Inc.
Suite 1100
1911 North Fort Myer Drive
Arlington VA 22209
(703) 525-0814

Personal Touch Corporation
Suite 290
4320 Stevens Creek Boulevard
San Jose CA 95129
(408) 248-8822

Philips UK
City House—420 London Road
Croydon ENG CR9 3QR
UNITED KINGDOM
(01)689-2166

Pioneer Communications of America
Industrial Laserdisc Division
600 East Crescent Ave.—Sherbrook Plaza
Upper Saddle River NJ 07458
(201) 327-6400

Pioneer Electronics USA
Consumer LaserDisc Division
2265 East 220th Street
Long Beach CA 90801
(213) 835-6177

Pioneer Electronics Europe
Keetberglaan 1
2740 Beveren
BELGIUM
(03)775-2808

Poseidon Systems
1898 Flatiron Court
Boulder CO 80301
(303) 449-4999

Professional Training Systems Inc.
500 Tech Parkway
Atlanta GA 30313
(404) 872-9700

Regency Systems Inc.
3200 Farber Drive
PO Box 3578
Champaign IL 61821
(217) 398-8067

RGB Dynamics
419 Wakara Way
Salt Lake City UT 84108
(801) 584-2550

Scion Corporation
12310 Pinecrest Boulevard
Reston VA 22091
(703) 476-6100

Schrello Direct Marketing
Suite 201—555 East Ocean Boulevard
PO Box 1610
Long Beach CA 90801
(213) 437-2230

Selection Systems Inc.
Suite 201
2731 77th Avenue SE
Mercer Island WA 98040
(206) 236-2700

SETS
Suite C4
4405 Vineland Road
Orlando FL 32811
(305) 422-7444

Sigma Electronics Inc.
1830 State Street
East Petersburg PA 17520
(717) 569-2681

**Software and Education
   Associates**
PO Box 99416
San Diego CA 92109
(619) 270-4570

Sony Intelligent Video Systems
One Sony Drive
Park Ridge NJ 07656
(201) 930-6000

Spectrum Training Systems
50 Salem Street
Lynnfield MA 01940
(617) 245-8500

Strategic Management Group Inc.
3624 Market Street
Philadelphia PA 19104
(215) 387-4000

Superior Training Systems
114 State Street
Boston MA 02109
(617) 523-4040

Symtec Inc.
15933 West Eight Mile Road
Detroit MI 48235
(313) 272-2950

Synetix Inc.
10635 NE 38th Place
Kirkland WA 98033
(206) 828-4884

Systems Impact Inc.
Suite 203
4400 MacArthur Boulevard NW
Washington DC 20007
(202) 342-9369

TEAC
7733 Telegraph Road
Montebello CA 90640
(213) 726-0303

Technovision Inc.
1087 Meyerside Drive—Unit 14
Mississauga ONT L5T 1M5
CANADA
(416) 671-8788

Teletape Video Ltd.
12 Golden Square
London ENG W1R 3AF
UNITED KINGDOM
(01)434-3311

Thomas Communications Ltd.
3845 Dutch Village Road
Halifax NOV B3L 4H9
CANADA
(902) 453-0040

Touch Technologies
9990 Mesa Rim Road
San Diego CA 92121
(619) 455-7404

University Associates Inc.
8517 Production Avenue
San Diego CA 92121
(619) 578-5900

University of Delaware
Office of Instructional Technology
Newark DE 19716
(302) 451-8161

University of Iowa
CAI Resource Lab 100
Weeg Computing Center
Iowa City IA 52242
(319) 335-5470

University of Washington
Health Science Center for Educational
Resources—T252-HSB-SB56
Seattle WA 98195
(206) 545-1186

University of West Florida
Office for Interactive Tech Training
11000 University Parkway
Pensacola FL 32514
(904) 474-2378

US Army—EIDS
Army Communicative Systems
PO Box 4337
Fort Eustis VA 23604
(804) 878-4210

US Video
Suite 140
900 Winderly Place
Maitland FL 32751
(305) 875-0800

Valiant IMC
195 Bon Homme Avenue
Hackensack NJ 07601

Vehiculum Corporation
PO Box 1785
Mathews NC 28106
(704) 376-1245

V-Graph Inc.
Box 105
Westtown PA 19395
(215) 399-1521

Video Associates Labs
4926 Spicewood Springs Road
Austin TX 78759
(512) 346-5781

Videodem Product Information Systems
East Wing—Springfield House
High Tech Center—Hyde Terrace
Leeds ENG LS2 9LU, U.K.
(0532)455-883

Visage
12 Michigan Drive
Natick MA 01768
(617) 655-1503

Visual Database Systems
614 Bean Creek Road
Scotts Valley CA 95066-3314
(408) 438-8396

WICAT Systems Inc.
1875 South State Street
PO Box 537
Orem UT 84057
(801) 224-6400

APPENDIX B: SELECTED INDUSTRIAL AND
TRAINING LASER VIDEODISCS

Videodisc Title/Producer	Player and Disc Type

Bio Sci Video Disc   All, Level I
    Videodiscovery, Inc.
    P.O. Box 85878
    Seattle, WA 98145

Cardiopulmonary Resuscitation (CPR)   Sony, Level III
    Interact, Inc.
    Chairman, Daniel Cassidy
    603 NE 27th Street
    Oklahoma City, OK 73105

The Challenges of Glaucoma   Pioneer, Level II
    Merck Sharp & Dohme
    126 E. Lincoln Avenue, Bldg. 33
    Rahway, NJ 07065

College U.S.A.   All, Level I
    Info-Disc Corp.
    451 Hungerford Drive
    Rockville, MD 20850

Core Concepts in Math and Science   All, Level I
    Systems Impact
    1220 Bank Street, NW
    Washington, D.C. 20007

Diagnosis and Management of a Pulmonary Problem   Pioneer, Level II
    Videodisc Design/Production Group
    KUON-TV
    P.O. Box 83111
    Lincoln, NE 68501

Videodisc Title/Producer	Player and Disc Type
Energy Transformations Featuring the Bicycle (Physics Instruction)	All, Level I
Studies in Motion (Physics Instruction) Great Plains National ITV Library Box 80669 Lincoln, NE 68501 or The Annenberg/CPB Project 1111 Sixteenth Street, NW Washington, D.C. 20036	
Evidence Objections Center for Computer Assisted Legal Instruction 229 19th Avenue South Minneapolis, MN 55455	Pioneer, Level III
Fair Employment Practices BNA Communications, Inc. 9401 Decoverly Hall Road Rockville, MD 20850	Pioneer, Level II
Gardening at Home Xerox Publishing Group 1 Pickwick Plaza P.O. Box 6710 Greenwich, CT 06836	All, Level I
Interactive Videodisc Instruction in Music Office of Computer-based Instruction Video Music Series Academy and Main Streets University of Delaware Newark, DE 19716	All, Level I
Let's Bowl *Omelettes* *Quiche–Chinese* *Lack of Self-Confidence* *Joy of Family* *Karate* *Finance* *Photography 1,2,3,4* *Investments* INEDCO 8925 N Meridan, #100 Indianapolis, IN 46260	Pioneer, Level I

Videodisc Title/Producer	Player and Disc Type

Math Assessment
(Bilingual grade 1–3 math skills assessment.
No reading skills required.)
    Systems Impact, Inc.
    2048 North 1200 East
    Logan, UT 84321

All, Levels I and III
Pioneer, Level II

PatSEARCH™
    Pergamon
    1340 Old Chain Bridge Road
    McLean, VA 22101

Pioneer, Level III

Physics and Automobile Collisions:
An Instructional Videodisc for
Laboratory Measurements
    Kansas State University
    Dept. of Physics
    Manhattan, KS 66506

Pioneer, Level III

Vincent/Van Gogh Revisited
    North American Philips
    100 East 42nd Street
    New York, NY 10017

All, Level I

Voyager
Apollo
Shuttle
Space Age
The Sun
    Video Vision Associates, Ltd.
    7 Waverly Place
    Madison, NJ 07940

All, Level I

APPENDIX C: INDUSTRY RESOURCES

3M
Optical Recording Project
223-5S 3M Center
St. Paul, MN 55144

McDonnell Douglas Electronics Co.
Box 426
St. Charles, MS 63302

Motivation Media, Inc.
1245 Milwaukee Blvd.
Glenview, IL 60025

New Orient Media, Inc.
Communications Building
2nd and Main St.
Box 333
West Dundee, IL 60118

Optical Disc Corp.
17517-H Fabrica Way
Cerritos, CA 90701

Panasonic Industrial Division
Matsushita Electronics Corp. of America
1 Panasonic Way
Secaucus, NJ 07094

Pioneer Industrial Video
5150 E Pacific Coast Hwy. Suite 300
Long Beach, CA 90804

Sony Video Communications
Sony Drive
Park Ridge, NJ 07656

Videodisc & Optical Disc-Videodisc
      Projects Directory
Meckler Publishing
11 Ferry Lane West
Westport, CT 06880

Videodisc Design/Production Group
KUON-TV/University of Nebraska-Lincoln
P.O. Box 83111
Lincoln, NE 68501-3111

WICAT Systems, Inc.
1875 S. State
P.O. Box 539
Orem, UT 84057

APPENDIX D: INFORMATION SOURCES

Books

Andriole, Peter J. *Interactive Computer-Based Systems.* Princeton, NJ: Petrocelli Books, 1983.

Daynes, Rod and Beverly Butler. *The Videodisc Book—A Guide and Directory.* New York: John Wiley and Sons, 1984.

DeBloois, M. *Videodisc/Computer Courseware Design.* Englewood Cliffs, NJ: Educational Technology Publications, 1982.

Ditlea, Steve, ed. *Digital Deli.* New York: Workman Publishing, 1984.

Iuppa, Nicholas V. *A Practical Guide to Interactive Video Design.* White Plains, NY: Knowledge Industry Publications, Inc., 1984.

Siegel, Efrem, Mark Schubin and Paul F. Merrill. *Videodiscs: The Technology, the Applications, and the Future.* New York: Van Nostrand Reinhold, 1980.

Periodicals

AV Video, Montage Publishing, Inc., 25550 Hawthorne Blvd., Torrance, CA 90505

American Cinematographer Magazine, ASC Holding Corp., Box 2230, Los Angeles, CA 90078

Audiovideo International, Dempa Publications, Inc., 400 Madison Ave., New York, NY 10017

Byte—The Small Systems Journal, Byte Publications, 70 Main St., Peterborough, NH 03458

CD-I News, Emerging Technologies and Publications, LINK Resources Corp., 79 Fifth Ave., New York, NY 10003

CD-ROM Review, 80 Elm St., Peterborough, NH 03458

Creative Computing Magazine, Ziff-Davis Publishing Co., One Park Ave., New York, NY 10016

Educational and Industrial Television, Tepfer Publishing Co., Inc., 51 Sugar Hollow Rd., Danbury, CT 06810

Educational Technology Systems, Society for Applied Learning Technology (SALT), 50 Culpeper St., Warrenton, VA 22186

International Television, Business Publications Division, Ziff-Davis Publishing Co., One Park Ave., New York, NY 10016

Optical Memory Newsletter, P.O. Box 14817, San Francisco, CA 94114

Popular Computing, McGraw-Hill Information Systems Co., Byte Publications, 70 Main St., Peterborough, NH 03458

Technological Horizons in Education Journal, Information Synergy, Inc., P.O. Box 992, Acton, MA 01720

Video Disc Monitor, Future Systems, Inc., P.O. Box 26, Falls Church, VA 02246 (also distributes *Interactive Videodisc Directory*)

Video Manager, Knowledge Industry Publications, Inc., 701 Westchester Ave., White Plains, NY 10604

Video Systems, Intertec Publishing Corp., 9221 Quivira Rd., P.O. Box 12901, Overland Park, KS 66212

Videodisc Design/Production Group News, P.O. Box 83111, Lincoln, NB 68501

Videodisc News, P.O. Box 6302, Arlington, VA 22206

Videodisc and Optical Disc, Meckler Publishing, 520 Riverside Ave., Westport, CT 06880

Videography, Media Horizons, Inc., 50 West 23rd St., New York, NY 10010

VideoPro, P.O. Box 1383, Pittsfield, MA 01202

INDEX

ABOUT THE AUTHOR

Gerald Souter, a video producer/director at Motivation Media in Glenview, IL produced, wrote and directed a multi-screen videodisc presentation for Nissan Motors' 1987 auto show exhibit. Formerly with Motorola Communications and Electronics, Inc., he has written, produced and directed numerous video, film and AV projects. Awards for his work include: the Gold Camera from the U.S. Industrial Film Festival for his videodisc, *Emergency, Emergency* (Motorola Communications), the Bronze Award from the Association for Multi-Image for *Motorola Joe,* a slide-film animation and a Tower Award for his *National Office Products Association* Video Wall (Wilson Jones). Gerald Souter also lectures on new communications media to corporate, professional, educational and community organizations.